AF371476

APPLIED TISSUE ENGINEERING

Edited by **Minoru Ueda**

Applied Tissue Engineering
Edited by Minoru Ueda

Published by InTech
Janeza Trdine 9, 51000 Rijeka, Croatia

Publishing Process Manager Jelena Marusic

Technical Editor Teodora Smiljanic

Cover Designer InTech Design Team

Additional hard copies can be obtained from orders@intechopen.com

Applied Tissue Engineering, Edited by Minoru Ueda
p. cm.
ISBN 978-953-307-689-8

Meet the editor

Dr. Minoru Ueda received a DDS degree from Tokyo Medical and Dental University in 1978 and a PhD degree in medical science from Nagoya University in 1982. He was appointed Professor and Chair of Oral and Maxillofacial Surgery at Nagoya University School of Medicine in 1994, also Visiting Professor of Stem Cell Engineering at Institute of Medical Science, University of Tokyo in 2003 and Korea University Ansan Hospital, Korea in 2011. He became Chairman of scientific advisory board at Japan Tissue Engineering, Inc. which was the first tissue engineering company in 1999 to commercialize the cultured skin. He received the "2004 President Award" from the Science Council of Japan for his contribution to the development of tissue engineered skin and cartilage. He is former President at Japanese Tissue Engineering Society, Asian Tissue Engineering Society, and Vice-President at Tissue Engineering International & Regenerative Medicine Society.

Contents

Preface

Tissue engineering, which aims at regenerating new tissues, as well as substituting lost organs by making use of autogenic or allogenic cells in combination with biomaterials, is an emerging biomedical engineering field. There are several driving forces that presently make tissue engineering very challenging and important: 1) the limitations in biological functions of current artificial tissues and organs made from man-made materials alone, 2) the shortage of donor tissue and organs for organs transplantation, 3) recent remarkable advances in regeneration mechanisms made by molecular biologists, as well as 4) achievements in modern biotechnology for large-scale tissue culture and growth factor production.

The idea of tissue engineering is not quite as new as it seems. The Nobel Laureate Alexis Carrel performed seminal work in the early 1900s that paved the way for today's tissue engineers. Carrel even caught the imagination of the pilot Charles Lindbergh. After his historic first solo flight across the Atlantic Ocean, Lindbergh worked with Carrel at the Rockefeller Institute in New York, with the goal of maintaining viable tissue and organs in vitro for subsequent implantation in vivo.

The earliest attempts at engineering tissue were carried out in skin by Bell, Yannas, and Green at the Massachusetts Institute of Technology in the late 1970s, early 1980s. Dr. Iannas Yannas, collaborated in studies in both the laboratory and in humans to generate a tissue-engineered skin substitute using a collagen matrix to support the growth of dermal fibroblasts. Dr. Eugine Bell later transferred sheets of keratocyte with fibroblasts referring to them as contracted collagen gels. From a more clinical aspect, one of the most exciting advances in culturing skin epithelium was carried out by Dr. Howard Green and it has been the ability to use these cells in treating patients with burns and other skin disorders, and this area is also covered in this book. The technique of culturing human skin epithelium in defined media and the exciting possibility to multiply epithelial cells up to ten thousand fold the amount of the original skin sample has been available for more than three decades by now. Many scientific pioneers in the field of biology, surgery and other disciplines have added their efforts with enormous creativity and ambition to further improve this method. The introduction of this technique into burn treatment in the early eighties of the past century was followed by initial enthusiasm about the prospect of saving the lives of many extensively burned patients. Our early success in Nagoya experiences have started by following the Green's artificial skin. Because he kindly provided the 3T3-J2 cell line and its supported our research works at a huge.

Tissue engineering was catapulted to the forefront of public awareness with the airing of a BBC broadcast on the potential of tissue-engineered cartilage using images of the now-infamous"mouse with the human ear", fondly referred to as auriculosaurus, from the laboratory of Dr. Charles Vacanti at University of Massachusetts Medical Center. This has become known as tissue engineering were all based on the same premise, that new functional replacement tissue could be generated from living cells seeded onto appropriately configured scaffoldings. In the example of cartilage, viable chondrocytes were seeded onto porous polymer fibers and configured in the shape of the desired tissue. Other potential applications of tissue engineering include the replacement of worn and poorly functioning tissues as exemplified by replacement of small caliber arteries, veins, bone and cartilage; replacement of the bladder, muscle and nerve tube; and restoration of cells to produce necessary enzymes, hormones, and other bioactive secretory products, such as salivary gland.

In spite of significant scientific progress in tissue engineering, there are few examples of human application. Two potential explanations for this may be 1) problems associated with "scale up" and 2) cell death associated with implantation. Large numbers of cells are needed to generate relatively small volumes of tissues. These problems are always associated with the human application of tissue engineering concept. Fortunately, dental field is the advantageous field because the volume of the newly formed tissue to be needed is relatively small compared with other field of medicine. However, to ultimately be effective in humans, tissue engineering must generate relatively large volume of tissue, starting with the very few cells. Cell implantation and its associated vascular disruption result in a relatively hypoxic environment and cell death. The potential for different cell types to be expanded in vitro and survive a relatively hostile environment at the time of implantation is now being explored. To be effective, cells should be easily procured, be effectively expanded in vitro, survive the initial implantation, be accepted as self, function normally, and not become malignant.

Embryonic stem cells (ES cells) and induced pluripotent stem cells (iPS cells) may have very similar potentials and risk to develop into the different cellular elements necessary for effective tissue regeneration. ES cells have been postulated to retain a greater ability to produce healthier tissue despite its ethical issues. At the point in time, there is little evidence that iPS cells can be consistently driven to form only the cell type needed for the tissue to be engineered without any risk of carcinogenicity. Being derived from autogenic cells, iPS cells have the associated problem of malignancy.

It is my belief that some forms of tissue-specific adult stem cells is the most hopeful cell at this moment for clinical use because it may represent Mother Nature's repair cells. Such cells are potentially present within all of the tissue of the body and may remain dormant until they are activated in response to tissue injury. Initially, the chemical environment at the site of any injury is very hostile. These adult stem cells, having a low oxygen requirement, appear to have the ability to survive this environment. When adequate numbers of cells have been achieved by multiplication, they are then programmed to mature and repair tissue damage of a certain magnitude. If this is the case, with the development of appropriate tissue-specific scaffolds and the use of the optimal cell type, I believe that physicians and scientists will ultimately be able to repair or replace any tissue in the human body that is injured or damaged as a result of disease or trauma. Studies involving the use of stem cells and mature cells, in combination with genetic manipulation and determination of the efficacy cellular delivery systems and scaffoldings, should be enable rapid progression to human treatments. It is my belief that exploring the use of appropriate vehicles and cell types will ultimately lead to resolution of stroke symptoms, such as paralysis, and may help reverse symptoms associated with such central nervous system diseases as Parkinson's disease and Alzheimer's disease. Along this stream we have been studying the usefulness of dental pulp stem cell as a new cell source for the central nerve regeneration.

This book was edited by collecting all the achievement performed in the laboratory of oral and maxillofacial surgery and it brings together the specific experiences of the scientific community in these experiences of our scientific community in this field as well as the clinical experiences of the most renowned experts in the fields from all over Nagoya University. The editors are especially proud of bringing together the leading biologists and material scientists together with dentist, plastic surgeons, cardiovascular surgery and doctors of all specialties from all department of the medical school of Nagoya University. Taken together, this unique collection of world-wide expert achievement and experiences represents the current spectrum of possibilities in tissue engineered substitution.

Minoru Ueda, DDS, PhD
Department of Oral & Maxillofacial Surgery,
Nagoya University Graduate School of Medicine

Acknowledgments

I would most importantly like to thank my wife, Hiroe and my son, Kouichiro, who supported me privately and officially. Without their encouragement these efforts would never have "gotten off the ground". I also would like to express my sincere gratitude to all of members of the Nagoya University who contribute as authors and subscribers to this book, edition for education and research of Micro-Nano Mechatronics, Nagoya University Global COE Program.

Contributors

Akihiro Abe, MD, PhD
Associate Professor
Department of Hematology
School of Medicine
Fujita Health University
Toyoake, Japan

Mika Aoki
Research Specialist
Institute of Molecular
Embryology and Genetics
Kumamoto University
Kumamoto, Japan

Katsumi Ebisawa, MD, PhD
Assistant Professor
Department of Plastic and
Reconstructive Surgery
Graduate School of Medicine
Nagoya University
Nagoya, Japan

Nobuhiko Emi, MD, PhD
Professor
Department of Hematology
School of Medicine
Fujita Health University
Toyoake, Japan

Atsushi Fujimoto, PhD
Director
TEI Corporation
Nagoya, Japan

Ken-Ichiro Hata, DDS, PhD
Director
Japan Tissue Engineering Corporation
Gamagori, Japan

Masao Hayashida, PhD
Former Researcher
Department of Biotechnology
School of Engineering
Nagoya University
Nagoya, Japan

Hideharu Hibi, DDS, PhD
Associate Professor
Department of Oral and
Maxillofacial Surgery
Graduate School of Medicine
Nagoya University
Nagoya, Japan

Yoshitaka Hibino, DDS, PhD
Lecturer
Department of Oral and
Maxillofacial Surgery
Graduate School of Medicine
Nagoya University
Nagoya, Japan

Hiroyuki Honda, PhD
Professor
Department of Biotechnology
School of Engineering
Nagoya University
Nagoya, Japan

Masaki J. Honda, DDS, PhD
Associate Professor
Second Department of Anatomy
Graduate School of Dentistry
Nihon University
Tokyo, Japan

Kunio Horie, DDS, PhD
Former Researcher
Department of Oral and
Maxillofacial Surgery
Graduate School of Medicine
Nagoya University
Nagoya, Japan

Akira Ito, PhD
Associate Professor
Department of Chemical
Engineering
Faculty of Engineering
Kyushu University
Fukuoka, Japan

Kenji Ito, DDS, PhD
Former Assistant Professor
Department of Oral and
Maxillofacial Surgery
Graduate School of Medicine
Nagoya University
Nagoya, Japan

Hideaki Kagami, DDS, PhD
Associate Professor
Division of Molecular Therapy
Advanced Clinical Research
Center
Institute of Medical Science
Tokyo University
Tokyo, Japan

Koji Kimata, MD, PhD
Director
Research Complex for
Medicine Frontiers
Aichi Medical University
Nagakute, Aichi, Japan

Kazuhiko Kinoshita, DDS, PhD
Clinical Staff
Department of Oral and
Maxillofacial Surgery
Graduate School of Medicine
Nagoya University
Nagoya, Japan

Kazuhiro Kito, MD, PhD
Former Researcher
Department of Ophthalmology
Graduate School of Medicine
Nagoya University
Nagoya, Japan

Chiaki Kobayashi, MD, PhD
Former Researcher
Department of Ophthalmology
Graduate School of Medicine
Nagoya University
Nagoya, Japan

Takeshi Kobayashi, PhD
Professor
Department of Biological Chemistry
College of Bioscience and Biotechnology
Chubu University
Kasugai, Japan

Tetsuhito Kojima, MD, PhD
Professor
Department of Medical Technology
School of Health Sciences
Nagoya University
Nagoya, Japan

Takahisa Kondo, MD, PhD
Lecturer
Department of Hematology
Graduate School of Medicine
Nagoya University
Nagoya, Japan

Junji Mase, DDS, PhD
Assistant Professor
Department of Public Health
School of Medicine
Fujita Health University
Toyoake, Japan

Hiroshi Matsunuma, MD, PhD
Former Researcher
Department of Urology
Graduate School of Medicine
Nagoya University
Nagoya, Japan

Yoshihisa Miyata, MD, PhD
Former Researcher
Department of Dermatology
Graduate School of Medicine
Nagoya University
Nagoya, Japan

Daiki Mizuno, DDS, PhD
Former Researcher
Department of Oral and
Maxillofacial Surgery
Graduate School of Medicine
Nagoya University
Nagoya, Japan

Hirokazu Mizuno, DDS, PhD
Former Researcher
Department of Oral and
Maxillofacial Surgery
Graduate School of Medicine
Nagoya University
Nagoya, Japan

Tetsuro Nagasaka, MD, PhD
Professor
Department of Medical
Technology
School of Health Sciences
Nagoya University
Nagoya, Japan

Takahito Naiki, DDS
Former Researcher
Department of Oral and
Maxillofacial Surgery
Graduate School of Medicine
Nagoya University
Nagoya, Japan

Yuji Narita, MD, PhD
Lecturer
Department of Cardiothoracic
Surgery
Graduate School of Medicine
Nagoya University
Nagoya, Japan

Atsushi Niimi, DDS, PhD
Lecturer
Department of Oral and
Maxillofacial Surgery
Graduate School of Medicine
Nagoya University
Nagoya, Japan

Hiroaki Nishiguchi, DDS, PhD
Lecturer
Department of Oral and
Maxillofacial Surgery
Graduate School of Medicine
Nagoya University
Nagoya, Japan

Takayuki Ohara
Engineer
System Design Department
MRI system Division
Hitachi Medical Corporation
Kashiwa, Japan

Shinichi Ohshima, MD, PhD
Director
National Center of Geriatrics
and Gerontology
Obu, Japan

Kunihiko Okada, PhD
President
Japan Animal Regenerative
Medicine Corporation
Osaka, Japan

Yoshinari Ono, MD, PhD
Professor
Faculty of Health and
Medical Sciences
Aichi Shukutoku University
Nagakute, Aichi, Japan

Hiroshi Sagara, PhD
Assistant Professor
Medical Proteomics Laboratory
Institute of Medical Science
Tokyo University
Tokyo, Japan

Hidehiko Saito, MD, PhD
Professor Emeritus
First Department of Internal
Medicine
Nagoya University
Nagoya, Japan

Yoshinori Shinohara, DDS, PhD
Clinical Staff
Department of Fixed Prosthodontics
Faculty of Dental Science
Kyushu University
Fukuoka, Japan

Yasuo Sugimura, DDS, PhD
Former Researcher
Department of Oral and
Maxillofacial Surgery
Graduate School of Medicine
Nagoya University
Nagoya, Japan

Takayuki Sugito, DDS, PhD
Former Researcher
Department of Oral and
Maxillofacial Surgery
Graduate School of Medicine
Nagoya University
Nagoya, Japan

Yukio Sumi, DDS, PhD
Former Researcher
Department of Oral and
Maxillofacial Surgery
Graduate School of Medicine
Nagoya University
Nagoya, Japan

Yoshinori Sumita, DDS, PhD
Assistant Professor
Department of Regenerative
Oral Surgery
Graduate School of
Biomedical Sciences
Nagasaki University
Nagasaki, Japan

Isao Takahashi, PhD
Section Chief
Department of Research
Japan Red Cross Aichi Blood Center
Seto, Japan

Hiroko Terasaki, MD, PhD
Professor
Department of Ophthalmology
Graduate School of Medicine
Nagoya University
Nagoya, Japan

Yasushi Tomita, MD, PhD
Professor Emeritus
Department of Dermatology
Graduate School of Medicine
Nagoya University
Nagoya, Japan

Akiko Tonomura
Development Design Department
Ultrasound Systems Division
Hitachi Medical Corporation
Kashiwa, Japan

Shuhei Torii, MD, PhD
Professor
School of Nursing
Sugiyama Jogakuen University
Nagoya, Japan

Kazuhiro Toriyama, MD, PhD
Associate Professor
Department of Plastic and
Reconstructive Surgery
Graduate School of Medicine
Nagoya University
Nagoya, Japan

Shuhei Tsuchiya, DDS, PhD
Assistant Professor
Department of Oral and
Maxillofacial Surgery
Graduate School of Medicine
Nagoya University
Nagoya, Japan

Yuichi Ueda, MD, PhD
Professor
Department of Cardiothoracic Surgery
Graduate School of Medicine
Nagoya University
Nagoya, Japan

Kazutaka Usami, DDS, PhD
Former Researcher
Department of Oral and
Maxillofacial Surgery
Graduate School of Medicine
Nagoya University
Nagoya, Japan

Akihiko Usui, MD, PhD
Associate Professor
Department of Cardiothoracic Surgery
Graduate School of Medicine
Nagoya University
Nagoya, Japan

Hideto Watanabe, MD, PhD
Professor
Institute for Molecular
Science of Medicine
Aichi Medical University
Nagakute, Aichi, Japan

Yoichi Yamada, DDS, PhD
Assistant Professor
Department of Oral and
Maxillofacial Surgery
Graduate School of Medicine
Nagoya University
Nagoya, Japan

Introduction

Tissue engineering is the use of a combination of cells, engineering, materials, methods, and suitable biochemical and physiochemical factors to improve or replace biological functions. While most definitions of tissue engineering cover a broad range of applications, in practice the term is closely associated with applications that repair or replace portions of or whole tissues (i.e., bone, cartilage, blood vessels, skin). Often, the tissues involved require certain mechanical and structural properties for proper functioning. The term has also been applied to efforts to perform specific biochemical functions using cells within an artificially-created support system (e.g., an artificial pancreas, or a bioartificial liver). The term regenerative medicine is often used synonymously with tissue engineering, although applications involved in regenerative medicine place more emphasis on the use of stem cells to produce tissues.

A commonly applied definition of tissue engineering, as stated by Langer and Vacanti, is "an interdisciplinary field that applies the principles of engineering and life sciences toward the development of biological substitutes that restore, maintain, or improve tissue function or a whole organ" [1]. Tissue engineering has also been defined as "understanding the principles of tissue growth, and applying this to produce functional replacement tissue for clinical use" [2]. A further description goes on to say that an "underlying supposition of tissue engineering is that the employment of natural biology of the system will allow for greater success in developing therapeutic strategies aimed at the replacement, repair, maintenance, and/or enhancement of tissue function."

Powerful developments in the multidisciplinary field of tissue engineering have yielded a novel set of tissue replacement parts and implementation strategies. Scientific advances in biomaterials, stem cells, growth and differentiation factors, and biomimetic environments have created unique opportunities to fabricate tissues in the laboratory from combinations of engineered extracellular matrices ("scaffolds"), cells, and biologically active molecules. Among the major challenges now facing tissue engineering is the need for more complex functionality, as well as both functional and biomechanical stability *in vitro*-grown tissues destined for transplantation. The continued success of tissue engineering, and the eventual development of true human replacement parts, will grow from the convergence of engineering and basic research advances in tissue, matrix, growth factor, stem cell, and developmental biology, as well as material science and bioinformatics. In 2003, the NSF published a report entitled "The Emergence of Tissue Engineering as a Research Field" [3], which gives a thorough description of the history of this field.

Cells

Cells are often categorized by their source:

- **Autologous** cells are obtained from the same individual to which they will be reimplanted. Autologous cells have the fewest problems with rejection and pathogen transmission, however in some cases might not be available. For example, in genetic disease suitable autologous cells are not available. Also very ill or elderly persons, as well as patients suffering from severe burns, may not have sufficient quantities of autologous cells to establish useful cell lines. Moreover since this category of cells needs to be harvested from the patient, there are also some concerns related to the necessity of performing such surgical operations that might lead to donor site infection or chronic pain. Autologous cells also must be cultured from samples before they can be used: this takes time, so autologous solutions may not be very quick. Recently, there has been a trend towards the use of mesenchymal stem cells from bone marrow and fat. These cells can differentiate into a variety of tissue types, including bone, cartilage, fat, and nerve. A large number of cells can be easily and quickly isolated from fat, thus opening the potential for large numbers of cells to be rapidly and readily obtained. Several companies have been established to capitalize on this technology, the most successful at this time being Cytori Therapeutics.

- **Allogenic** cells come from the body of a donor of the same species. While there are some ethical constraints to the use of human cells for *in vitro* studies, the employment of dermal fibroblasts from human foreskin has been demonstrated to be immunologically safe and thus a viable choice for tissue engineering of skin.

- **Xenogenic** cells are those isolated from individuals of another species. In particular animal cells have been used quite extensively in experiments aimed at the construction of cardiovascular implants.

- **Syngenic** or **isogenic** cells are isolated from genetically identical organisms, such as twins, clones, or highly inbred research animal models.

- **Primary** cells are from an organism.

- **Secondary** cells are provided by a cell bank.

- **Stem cells** are undifferentiated cells with the ability to divide in culture and give rise to different forms of specialized cells. According to their source, stem cells are divided into "adult" and "embryonic" stem cells, the first group being multipotent and the latter mostly pluripotent; some cells are totipotent, in the earliest stages of the embryo. While there is still a large ethical debate related with the use of embryonic stem cells, it is thought that stem cells may be useful for the repair of diseased or damaged tissues, or may be used to grow new organs.

- **Induced pluripotent stem (iPS) cells** are a type of pluripotent stem cell artificially derived from a non-pluripotent cell, typically an adult somatic cell, by inducing a "forced" expression of certain genes. Induced pluripotent stem cells are believed to be identical to natural pluripotent stem cells, such as embryonic stem cells in many respects, such as the expression of certain stem cell genes and proteins, chromatin methylation patterns, doubling time, embryoid body formation, teratoma formation, viable chimera formation, potency, and differentiability. However, the full extent of their relation to natural pluripotent stem cells is still being assessed. Induced pluripotent stem (iPS) cells were first produced in 2006 from mouse cells and in 2007 from human cells. This has been cited as an important advancement in stem cell research, as it may allow researchers to obtain pluripotent stem cells, which are important in research and potentially have therapeutic uses, without the controversial use of embryos.

Scaffolds

Cells are often implanted or "seeded" into an artificial structure capable of supporting three-dimensional tissue formation. These structures, typically called scaffolds, are often critical, both *ex vivo* as well as *in vivo*, to recapitulating the *in vivo* milieu and allowing cells to influence their own microenvironments. Scaffolds usually serve at least one of the following purposes:

- Allow cell attachment and migration
- Deliver and retain cells and biochemical factors
- Enable diffusion of vital cell nutrients and expressed products
- Exert certain mechanical and biological influences to modify the behavior of the cell phase

Carbon nanotubes are among the numerous candidates for tissue engineering scaffolds since they are biocompatible, resistant to biodegradation and can be functionalized with biomolecules. However, the possibility of toxicity with non-biodegradable nanomaterials is not fully understood.

To achieve the goal of tissue reconstruction, scaffolds must meet some specific requirements. High porosity and adequate pore size are necessary to facilitate cell seeding and diffusion throughout the whole structure of both cells and nutrients. Biodegradability is often an essential factor since scaffolds should preferably be absorbed by the surrounding tissues without the necessity of surgical removal. The rate at which degradation occurs has to coincide as much as possible with the rate of tissue formation: this means that while cells are fabricating their own natural matrix structure around themselves, the scaffold is able to provide structural integrity within the body and eventually break down, leaving neo tissue, the newly formed tissue which will take over the mechanical load. Injectability is also important for clinical uses.

Growth Factors

The term growth factor refers to a naturally occurring protein capable of stimulating cellular growth, proliferation and cellular differentiation. Growth factors are important for regulating a variety of cellular processes. Growth factors typically act as signaling molecules between cells. Examples are cytokines and hormones that bind to specific receptors on the surface of their target cells.

They often promote cell differentiation and maturation, which varies between growth factors. For example, bone morphogenic proteins stimulate bone cell differentiation, while fibroblast growth factors and vascular endothelial growth factors stimulate blood vessel differentiation (angiogenesis).

Classes of growth factors

Individual growth factor proteins tend to occur as members of larger families of structurally and evolutionarily related proteins. There are many families which are listed below:

- Bone morphogenetic proteins (BMPs)
- Epidermal growth factor (EGF)
- Erythropoietin (EPO)
- Fibroblast growth factor (FGF)
- Granulocyte-colony stimulating factor (G-CSF)

- Granulocyte-macrophage colony stimulating factor (GM-CSF)
- Growth differentiation factor-9 (GDF9)
- Hepatocyte growth factor (HGF)
- Insulin-like growth factor (IGF)
- Myostatin (GDF8)
- Nerve growth factor (NGF) and other neurotrophins
- Platelet-derived growth factor (PDGF)
- Thrombopoietin (TPO)
- Transforming growth factor alpha (TGF-α)
- Transforming growth factor beta (TGF-β)
- Vascular endothelial growth factor (VEGF)

Tissue engineering can be used to restore, maintain, or enhance tissues and organs. The potential impact of this field, however, is far reaching. Engineered tissues could reduce the need for organ replacement, and could greatly accelerate the development of new drugs that may cure patients, eliminating the need for organ transplants altogether. Scientists in the field of tissue engineering apply the principles of cell transplantation, material science, and bioengineering to construct biological substitutes that will restore and maintain normal function in diseased and injured tissues. Various organs and tissues are at different stages of development, with some already being used clinically like our "injectable bone" described below.

References

1. Langer R, Vacanti JP. Tissue engineering. Science. **260**: 920,1993
2. MacArthur BD, Oreffo ROC. Bridging the gap. Nature. **433**: 19,2005
3. http://www.nsf.gov/pubs/2004/nsf0450/start.htm

Chapter 1

Skin

In the field of plastic and reconstructive surgery, many clinical reports have demonstrated the utility of cultured epithelial sheets which have been successfully cultured since the first report of Rheinwald and Green.

Cultivation of human epidermal keratinocytes

Human diploid epidermis epidermal cells have been successfully grown in serial culture. To initiate colony formation, they require the presence of fibroblasts, but proliferation of fibroblasts must be controlled so that the epidermal cell population is not overgrown. Both conditions can be achieved by the use of lethally irradiated 3T3 cells at the correct density. When trypsinized human skin cells are plated together with the 3T3 cells, the growth of the human fibroblasts is largely suppressed, but epidermal cells grow from single cells into colonies. Each colony consists of keratinocytes ultimately forming a stratified squamous epithelium in which the dividing cells are confined to the lowest layer(s). Hydrocortisone is added to the medium, since in secondary and subsequent subcultures it makes the colony morphology more orderly and distinctive, and maintains proliferation at a slightly greater rate. Under these culture conditions, it is possible to isolate keratinocyte clones free of viable fibroblasts. Like human diploid fibroblasts, human diploid keratinocytes appear to have a finite culture lifetime. For 7 strains studied, the culture lifetime ranged from 20-50 cell generations. The plating efficiency of the epidermal cells taken directly from skin was usually 0.1-1.0%. On subsequent transfer of the cultures initiated from newborns, the plating efficiency rose to 10% or higher, but was most often in the range of 1-5% and dropped sharply toward the end of their culture life. The plating efficiency and culture lifetime were lower for keratinocytes of older persons.

Grafting of burns

The cells from a small piece of epidermis can be grown into a large number of cultured epithelia. Such epithelia, generated from autologous skin, were grafted onto full-thickness burn wounds in two patients. The cultured epithelia acquired an epidermal structure resembling that achieved with conventional split-thickness skin grafts, and survived for the period of observation (up to 8 months). Since the method of cultivation can generate large amounts of epithelium, the procedure is applicable to the grafting of large areas, as in severe burns (Fig. 1). However, using the method of epithelial cell culture employing 3T3 cells as a feeder layer, the epidermal cells possess a lower growth potential than normal cells such as fibroblasts. Therefore, approximately 3 weeks are required to fabricate an epidermal sheet. This makes it difficult to meet the requirements of emergency surgery. There have been some clinical trials of cultured epithelial sheet freezing storage in the field of plastic surgery [1, 2]. Thus, various freezing methods were considered to maintain the activity of the cultured epithelial sheets [3, 4]. These reports indicated that it is possible to store many cultured epithelial sheets long-term by using the freezing storage method and to meet the requirements of emergency surgery.

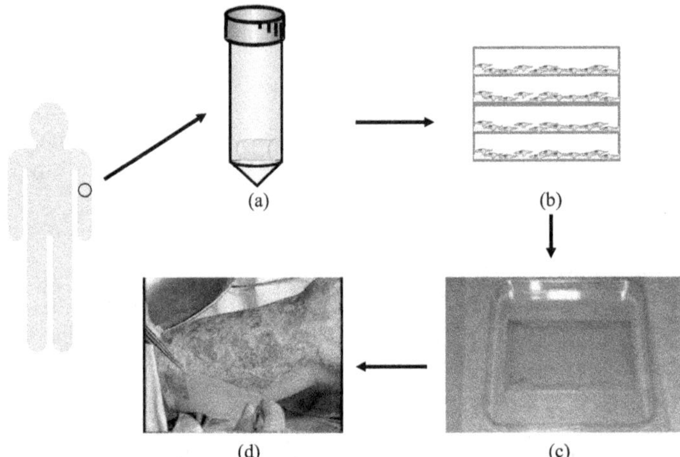

Fig. 1. Protocol of tissue engineered epithelium. (a) Extract skin tissue. (b) Cultured fibroblast on feeder layer. (c) Preparation of cultured skin. (d) Transplantation of cultured skin (From Ueda et al. 1995).

Effects of freezing storage

Numerous clinical reports have shown the utility of cultured epithelial grafting in the field of plastic and reconstructive surgery. Recently, freezing storage of the cultured epithelium has been tried and has successfully grafted after thawing. It is clinically convenient if it is possible for cultured epithelium to keep its normal structure and viability. However, few papers have described the structural changes in cultured epithelium after freezing storage. In the present study, the morphological changes and cell viability of cultured mucosal epithelial sheets after freezing were studied in comparison with cultured epidermal sheets. Furthermore, we discuss the effect of storage temperature and cryoprotectants. As a result, there were some structural changes such as vacuolar degeneration in the cultured mucosal sheets using dimethyl sulphoxide (DMSO) as a cryoprotectant. Such changes were more dearly observed at -80°C than at -196°C with DMSO. However, little morphological change was observed in both epithelial sheets cultured with glycerin. The cell viability analyzed by flow cytometry showed that more than 62% of the cells kept their viability after freezing storage. These results suggest that the optimum conditions of freezing for cultured epithelium were -196°C storage by slow cooling methods with glycerin as a cryoprotectant.

A telomere is a special base-sequence repeat at the end of a eukaryotic chromosome (TTAGGG repeat in humans) [5, 6]. Telomeres are required for protecting chromosomes against illegitimate fusion events, mediating chromosome location in the nucleus, and preventing the outermost end from being recognized as defective DNA [7, 8]. Thus, they function as important buffers to guarantee the stability and functionality of the chromosomes. Since a telomere does not replicate completely during cell division, it gradually shortens as cell division proceeds [9-11]. Over a single lifetime, up to 10 kb of telomeres are reduced gradually to around 5 kb after repeated cell division, resulting in cellular senescence [12]. From investigations using fibroblasts, telomere length has been proven to have a close correlation with the possible number of cell divisions [12]. Telomeres shorten during cell divisions without telomerase activity. Telomerase is a specialized ribonucleoprotein complex that directs the synthesis of telomeric repeats at chromosome ends [13]. Telomerase activity maintains telomeric DNA within relatively constant ranges, allowing cells, e.g. some neoplastic cells, to proliferate indefinitely [14, 15]. Although no telomerase activity is observed in most somatic cells, it has been detected in regenerating cells such as basal cell populations in skin epithelium [16-19]] Recently, the loss of telomerase activity has reportedly been associated with the replicative senescence of normal human oral and skin keratinocytes [16, 20]. Thus, the change in telomerase activity during long-term cultivation is of great interest, since it might be involved in the mechanisms underlying cellular senescence in cultured epidermis.

The telomere length and morphology
Cultured epidermis has been successfully used in clinical treatment such as burns and pigmentary disorders. Although the generation of wide cultured epidermis for clinical use may require repeated passages, especially for allografts, the effects of long-term cultivation on its quality and cell viability are not well known. To investigate the changes in morphology, telomere length, and telomerase activity during the passages of cultured epidermis and keratinocytes up to the passage limit, and to examine the usefulness of telomere length as a performance criterion for cultured epidermis. The keratinocytes obtained from five patients were used to generate cultured epidermis (Fig. 2). At the early passage and after cultivation up to the passage limit, morphology, telomere length and telomerase activity were investigated by using microscopes, southern blot analysis and telomeric repeat amplification protocol assay, respectively. The cultured cells started to show morphological changes when each passage was close to its limit and the cell sheets assumed an irregular stratification with various sizes of cytoplasm and nuclei. At the passage limit, the telomere length had decreased approximately 80-85%, and the average telomerase activity had declined under serum-free culture conditions. The results of this study showed the morphological change and telomere length reduction by long-term cultivation on cultured epidermis. Although the reduction in telomere length and telomerase activity may not be the major cause of the senescence, they could provide an useful information for the quality of the cultured epidermis. Dispase, a neutral protease from *Bacillus polymyxa*, is widely used to harvest multilayered keratinocyte sheets from culture dishes [16]. In clinical use, extensive washing to remove dispase from keratinocyte sheets is required before they can be applied to wounds, because residual dispase is harmful to the wounded site. In industrial production of keratinocyte sheets, this washing is laborious, and is a technological barrier to automation of the process. In the present study, we used ultralow-attachment plates with a surface composed of a covalently bound hydrogel layer that is hydrophilic and neutrally charged, to investigate whether keratinocytes could be harvested from the plate without enzymatic treatment after removing the magnet, because the keratinocytes did not adhere to the plate surface. The present results indicate that magnetic force and magnetite nanoparticles can be used to construct and harvest keratinocyte sheets.

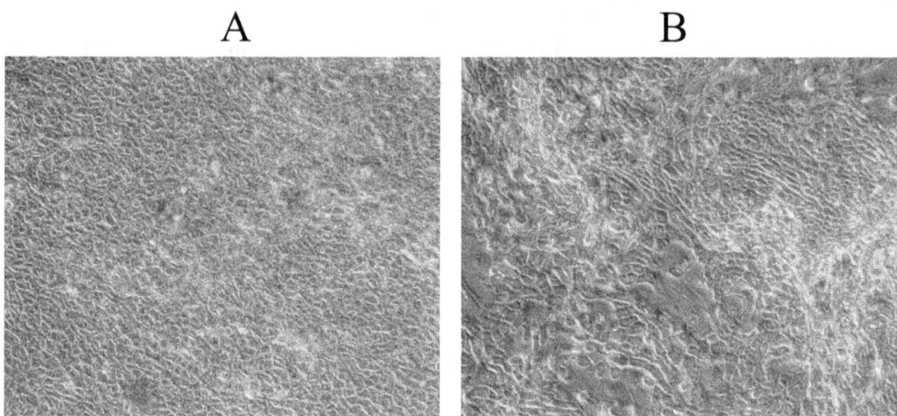

Fig. 2. A: The cultured epidermis in passage 1. During the first passage, the keratinocytes formed a sheet, and showing a uniform and typical cobble stone-like appearance. B: The cultured epidermis in passage 6. Around the terminal passage, the sheets displayed an uneven thickness with irregular cell shapes (From Miyata et al. 2004. Reprinted with permission).

Multilayered keratinocyte sheets using magnetite nanoparticles

Novel technologies to establish three-dimensional constructs are desired for tissue engineering. In the present study, magnetic force was used to construct multilayered keratinocyte sheets and harvest the sheets without enzymatic treatment. Our original magnetite cationic liposomes, which have a positive surface charge in order to improve adsorption, were taken up by human keratinocytes at a concentration of 33 pg of magnetite per cell. The magnetically labeled keratinocytes (2×10^6 cells, which corresponds to 5 times the confluent concentration against the culture area of 24-well plates, in order to produce 5-layered keratinocyte sheets) were seeded into a 24-well ultralow-attachment plate, the surface of which was composed of a covalently bound hydrogel layer that is hydrophilic and neutrally charged. A magnet (4000 G) was placed under the well, and the keratinocytes formed a 5-layered construct in low-calcium medium (calcium concentration, 0.15 mM) after 24 hours of culture. Subsequently, when the 5-layered keratinocytes were cultured in high-calcium medium (calcium concentration, 1.0 mM), keratinocytes further stratified, resulting in the formation of 10-layered epidermal sheets. When the magnet was removed, the sheets were detached from the bottom of the plates, and the sheets could be harvested with a magnet. These results suggest that this novel methodology using magnetite nanoparticles and magnetic force, which we have termed "magnetic force-based tissue engineering" (Mag-TE), is a promising approach for tissue engineering.

Mucosa

In the field of oral surgery, mucosal grafting has been carried out for reconstruction after tumor removal, preprosthetic surgery and so on. However, it is difficult to obtain enough oral mucosa for reconstruction of large defects, and in these cases, skin autografting has been employed. Skin grafting in the oral region has many disadvantages, such as keratinization, secretion and hair growth, and it is inevitably accompanied by patient discomfort. Furthermore, a skin graft is associated with a new defect at the donor site. To solve these problems, the use of cultured gingival epithelium was reported.

Formation of epithelial sheets

A cultured epithelial sheet can be formed from living mucosal cells *in vitro* and used as a graft material (Fig. 3). In this article, we describe our culturing methods for the preparation of mucosal epithelial sheets as well as the biological characteristics of these sheets compared with those of skin epithelial sheets. A cultured epithelial sheet has 5 to 8 cell layers and sufficient mechanical strength to be used as a graft material. It takes 12 days to form an epithelial sheet from small epithelial segments as compared with 14 days in the case of a skin epithelial sheet. Furthermore, viability of mucosal epithelial sheets was maintained for 30 days *in vitro* as opposed to 22 days for skin epithelial sheets. Based on the findings from an *in vitro* study, we applied this cultured mucosal epithelium to humans for reconstruction of skin and mucosal defects and succeeded in repairing the defects. This report also presents an overview of the problems relevant to the use of such methods.

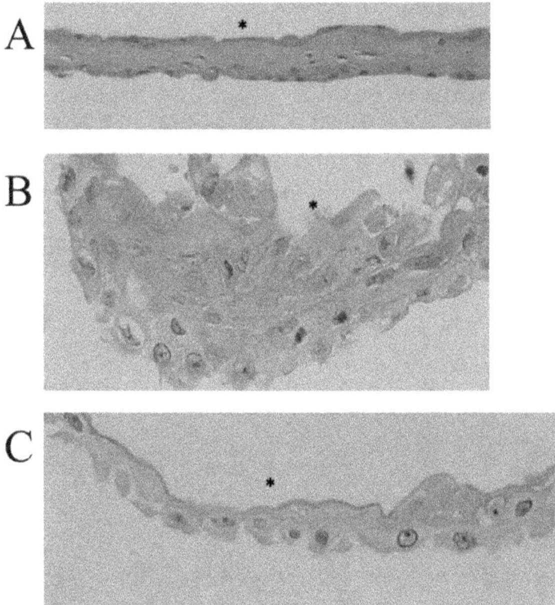

Fig. 3. A: The cultured epidermis in passage 2. The sheet showed 5 to 8 cell layers of even thickness including a single basal-cell layer. The polyhedral cells of stratum overlying the basal cells became flattened from the basal (plate) side toward to the surface (medium) side (indicated by an asterisk), and the cells in the surface layer showed an enucleation. B: The cultured epidermis in passage 6, 3 weeks after beginning of the culture. The cell sheets presented an irregular stratification with various sizes of cytoplasm and nuclei. C: When the cells reached the full passage limit, most areas of the sheet presented a single cell layer and only partial stratification (From Miyata et al. 2004. Reprinted with permission).

The characteristics of cultured mucosal cell sheet

The characteristics of cultured mucosal cells from the oral mucosa were investigated and compared with those of cultured epidermal cells. Total cell counts showed that mucosal cells possessed greater proliferating ability than epidermal cells. The results of 3(4,5-dimethylthiazole-2-yl)-2,5-diphenyltetrazolium bromide assay confirmed this observation and also suggested that the mucosal cells maintained biological activity longer than epidermal cells. The most important morphological characteristics of mucosal cells in culture were their low grade of differentiation. Interestingly, the epidermal cells showed enucleation and keratinization progressively during culture, whereas the mucosal cells showed no obvious enucleation when examined by light microscopy. Transmission electron microscopy showed a smaller number of desmosomes in cultured mucosal cells than epidermal cells. The results of this study reveal cultured mucosal cell sheets to be a possible material for grafting in addition to cultured epidermal cell sheets.

Transplantation of cultured mucosal epithelium

We investigated morphological changes after transplantation of cultured mucosal epithelium using a modified Barrandon's method (1988). Serially cultivated human mucosal epithelium was transplanted onto the reverse side of rectangular dorsal skin flaps in hairless mice. The morphological changes in the epithelium were studied using paraffin sections. The modified Barrandon's method used in this study has advantages such as minimum external trauma and less chance of infection. The cultured epithelium was taken within 1 week and gradually increased its epithelial thickness (Fig. 4). Keratinized epithelium arises after 3 weeks. At 4 weeks after grafting, the grafted epithelium comprised 7 to 10 cell layers. The structure of transplanted tissue, in conjunction with surrounding connective tissues, showed dermis-like features at day 7 after transplantation. From these results, it was confirmed that cultured mucosal epithelium could be successfully transplanted and its morphology was similar to that of normal mucosal tissue.

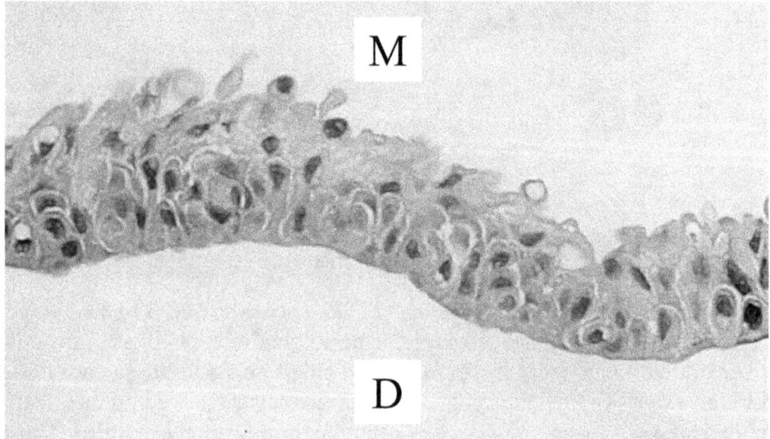

Fig. 4. Bright field photomicrographs of H-E stained sections of cultured mucosal epithelium of day 14 (× 250) (M: medium side, D: dish side) (From Sugimura et al. 1997. Reprinted with permission).

Grafting in the oral and maxillofacial region

Cultured epithelium has proven to be a good grafting material for skin defects. In our experience two kinds of epithelial cells, skin keratinocytes and mucosal cells, have been used to fabricate cultured epithelial sheets and autografted to the patients. Traumatic scars of the face were treated by cultured epidermal epithelium (CEE). The skin graft in the oral cavity was replaced by mucosa using cultured mucosal epithelium (CME). Also, the CME was applied to the skin defects at the donor sites of split-thickness skin grafts. Postsurgical follow-up showed good results. As a result, CME was useful in improving the biological environment around the abutments of dental implants, and it also promoted the re-epithelialization of skin defects. From our investigations, CEE/CME are promising treatment modalities which can reduce pain and speed up the healing process in burn patients. Therefore, cultured epithelium banks are worth establishing for autografting and allografting of skin/mucosal defects.

Peri-implant soft tissue management

In implant therapy, peri-implant soft tissue management through use of mucosal grafting or skin grafting is necessary in some patients who do not have enough attached gingiva around the abutment. However, limitation of donor site size is a problem for the mucosal graft, and the different characteristics of skin, such as hair growth, are disadvantages in treatment that involves the use of skin graft. On the other hand, cultured epithelium fabricated with living mucosal cells has proved to be a good grafting material for any kind of mucosal defect. In this study, we used cultured mucosal epithelium for soft tissue management in implant therapy. In the first surgical procedure of the implant therapy, a small segment of oral mucosa was sampled from a patient. The cultured epithelium was fabricated and then stored until it was grafted in the second surgery. Twelve cases in which patients underwent peri-implant soft tissue management through use of cultured mucosal epithelium for implant therapy are presented, and the usefulness of this technique in the making of attached gingiva is analyzed. From this study it was concluded that cultured mucosal epithelium can serve as a proper material for peri-implant soft tissue management.

Gene-modified mucosal epithelium

Human oral mucosal cells are attractive sites for tissue engineering because they are the most accessible cells in the body and easy to manipulate *in vitro*. They thus have possibilities for targeting by somatic gene therapy. We examined the efficiency of retrovirus-mediated gene transfer and the construction of mucosal epithelium *in vivo*. Human oral mucosal cells were transduced with a retroviral vector carrying the *lacZ* gene at high efficiency and constructed epithelium after G418 selection with 3T3 cells *in vitro*. The cultured oral mucosal epithelium membrane was then grafted onto immunodeficient mice. β-Gal expression was detected histochemically *in vivo* 5 weeks after grafting. Furthermore, we transduced factor IX cDNA into the mucosal epithelium membrane, and it was then transplanted into nude mice. Between 0.6 and 1.8 ng of human factor IX per milliliter was found in mouse plasma, and the production was continued for 23 days *in vivo*. These results confirmed that the oral mucosal epithelium is an ideal target tissue for gene therapy or tissue engineering.

The use of gingival fibroblasts for soft-tissue augmentation

Fibroblasts were obtained from the patient's buccal gingival tissue and maintained in DMEM plus 10% autologous human serum, and incubated at 37°C with 5% CO_2. Autologous patient serum was prepared from 100–150 ml of peripheral blood. The characteristics of the cultured cells were checked by immunofluorescent microscopy for known fibroblast markers. The cells were washed twice with sterile saline, and resuspended in sterile saline to a final concentration of 1.0×10^7 cells/ml. The cell suspension was stored in 1.0 ml syringes until use. All satisfactory assessments were performed by the patient, and the following grading scale was used at 3, 6 and 12 months after the first injection (4: Completely satisfied, 3: Satisfied, 2: No remarkable change observed, 1: Not satisfied, 0: Exacerbation). Several patients were treated with live gingival fibroblast injections. The population was 100% female. Mean age at the time of the first injection was 50.29 years. At 3-month follow-up, the satisfactory rate among fibroblast-treated patients was 3.0 in nasolabial and 3.00 in lip. At 6 months, the satisfactory rate increased to 3.29, whereas the lip increased relatively at 4.00. At 12 months, the rates were 3.36 versus 4.00. No serious adverse events considered related to the study treatment were reported. Our results indicate that autologous fibroblast injections may provide an acceptable treatment for patients, especially those desiring better aesthetic results. Our initial experience with the autologous gingival fibroblast injection process indicates that it is probably capable of producing ongoing improvements in perioral lip wrinkles without the hypersensitivity complications and harvesting challenges associated with other treatments (Figs. 5, 6).

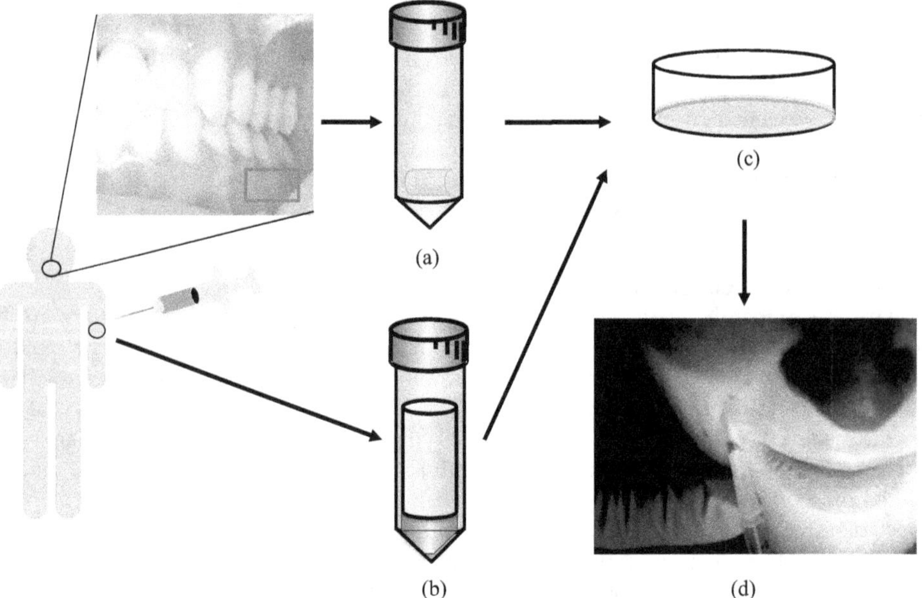

Fig. 5. Protocol of autologous fibroblast injections for soft-tissue augmentation. (a) Oral mucosa. (b) Autologous serum. (c) Cultured fibroblast with autologous serum. (d) Cell transplantation (From Ebisawa et al. 2008. Reprinted with permission).

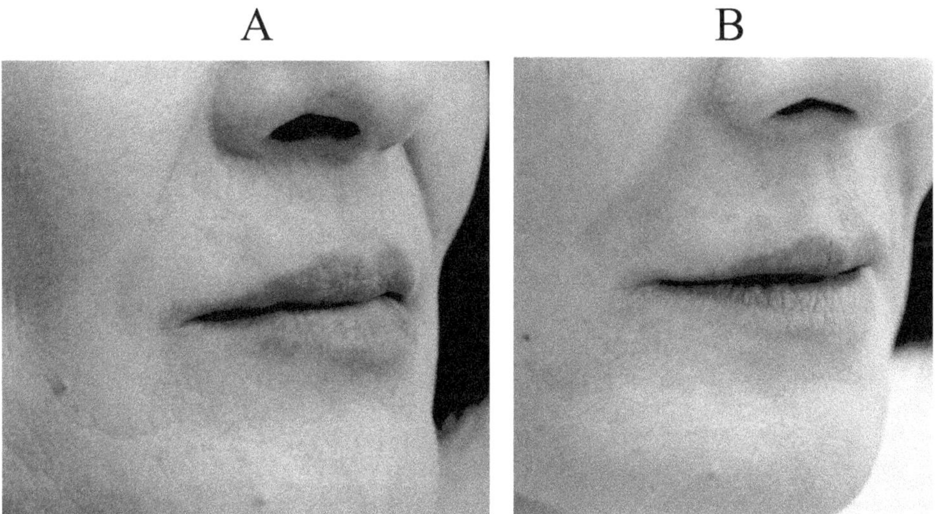

Fig. 6. A: Before treatment. B: Three years after treatment (From Ebisawa et al. 2008. Reprinted with permission).

References

1. De Luca M, Albanese E, Bondanza S. Multicentre experience in the treatment of burns with autologous and allogenic cultured epithelium, fresh or preserved in frozen state. Burns. **15**: 303,1989

2. Teepe RGC, Koebrugge EJ, Ponec M. Fresh versus cryopreserved cultured allografts for the treatment of chronic skin ulcers. Br J Dermatol. **122**: 81,1990

3. Hata K, Kagami H, Ueda M. An experimental study of cultured epithelial grafting using oral mucosal cells. Fabrication of cultured epithelium and its morphological analysis. J Jpn Stomatol Soc. **43**: 416,1994

4. Hata K, Kagami K, Ueda M. Matsuyama, M. The characteristics of cultured mucosal cell sheet as material for grafting; Comparison with cultured epidermal cell sheet. Ann Plast Surg. **34**: 530,1995

5. Blackburn EH, Gall JG. A tandemly repeated sequence at the termini of the extrachromosomal ribosomal RNA genes in tetrahymena. J Mol Biol. **120**: 33,1978

6. Moyzis RK, Buckingham JM, Cram LS. A highly conserved repetitive DNA sequence, $(TTAGGG)_n$, present at the telomeres of human chromosomes. Proc Natl Acad Sci USA. **85**: 6622,1988

7. Evans SK, Bertuch AA, Lundbald V. Telomeresand telomerase: at the end, it all comestogether. Trends Cell Bio. **19**: 329,1999

8. Zakian VA. Structure and function of telomeres. Annu Rev Genet. **23**: 579,1989

9. Olovnikov AM. Principle of marginotomy in template synthesis of polynucleotides. Dokl Akad Nauk Sssr. **201**: 1496,1971

10. Watson JD. Origin of concatameric T7 DNA. Nat New Biol. **239**: 197,1972

11. Harley CB, Futcher AB, Greder CW. Telomeres shorten during aging of human fibroblasts. Nature. **345**: 458,1990

12. Allsopp RC, Vaziri H, Patterson C. Telomere length predicts replicative capacity of human fibroblasts. Proc Natl Acad Sci USA. **89**: 10114,1992

13. Greider CW, Blackburn EH. A telomeric sequence in the RNA of tetrahymena telomerase required for telomere repeat synthesis. Nature. **337**: 331,1989

14. Harley CB. Telomere loss: mitotic clock or genetic time bomb? Mutat Res. **256**: 271,1991

15. Counter CM, Avilion AA, LeFeuvre CE. Telomere shortening associated with chromosome instability is arrested in immortal cells which express telomerase activity. EMBO J. **11**: 1921,1992

16. Yasumoto S, Kunimura C, Kikuchi K. Telomerase activity in normal human epithelial cells. Oncogene. **13**: 433,1996

17. Tayler RS, Ramirez RD, Ogoshi M. Detection of telomerase activity in malignant and non-malignant skin condition. J Invest Dermatol. **106**: 759,1996

18. Harle-Bachchor C, Boukamp P. Telomerase activity in the regenerative basal layer of epidermis in human skin and in immortal and carcinoma-derived skin keratinocytes. Proc Natl Acad Sci USA. **93**: 6476,1996

19. Kannan S, Tahara H, Yokozaki H. Telomerase activity in premalignant and malignant lesions of human oral mucosa. Cancer Epidemiol Biomarkers. **16**: 413,1997

20. Kang MK, Guo W, Park NH. Replicative senescence of normal human oral keratinocytes is associated with the loss of telomerase activity without shortening of telomeres. Cell Growth Differ. **9**: 85,1998

21. Matsui M, Miyasaka J, Hamada K. Influence of aging and cell senescence on telomerase activity in keratinocytes. J Dermatol Sci. **22**: 80,2000

21. Martin IC, Brown AE. Free vascularised fascial flap in oral cavity reconstruction. Head Neck. **16**: 45,1994

(Kito K, Kagami H, Kobayashi C, Terasaki H, Miyata Y, Okada K, Fujimoto A, Hata K, Tomita Y, Ito A, Hayashida M, Honda H, Kobayashi T, Sugimura Y, Torii S, Horie K, Hibino Y, Toriyama K, Sumi Y, Mizuno H, Niimi A, Emi N, Abe A, Takahashi I, Kojima T, Saito H, Ueda M)

Chapter 2

Cornea

Effects of cryopreservation against cultured corneal epithelial cell sheets

The recent development of tissue engineering [1, 2] has enabled the clinical application of cultured corneal epithelial cell sheets. In particular, the cultured corneal epithelial cell sheet is beneficial for the reconstruction of the ocular surface with limbal stem cell deficiencies [3], such as those with chemical and thermal injuries, ultraviolet and ionizing radiation injuries, Stevens-Johnson syndrome, and ocular cicatricial pemphigoid. Because the sheets can be generated from a small piece of limbal tissue that contains corneal epithelial stem cells [2], cultured corneal epithelial sheet transplantation is less invasive than autolimbal transplantation. However, the shortage of human donors is one of the major problems for the timely supply of cultured corneal epithelial cell sheets. Furthermore, the cell culture usually requires several weeks to generate stratified sheets [4] because the corneal epithelial cells possess a lower growth potential than other types of cells such as fibroblasts [1, 2]. The corneal epithelial cells tend to differentiate and lose their viability when the culture period is longer. To overcome these problems, cryopreservation should become an important option to preserve tissue-engineered corneal epithelium.

To date, various methods for the cryopreservation have been reported to maintain the viability of the cells for a long time without affecting the phenotype. The ability of glycerol to protect cells from freezing injury was discovered accidentally. Polge and coworkers found that glycerol can increase the survival rate of spermatozoa dramatically when added to a freezing medium [5]. Since then, cryopreservation techniques have been developed in various fields including blood cell storage, plant biology, and horticulture. Simultaneously, cryoprotective ability has been found in various chemicals. Among those chemicals that have been tested as cryoprotectants, glycerol and DMSO are most commonly used at present. Cryopreservation has also become an important technique in the field of tissue engineering because effective preservation procedures are required for a stable supply and the efficient transportation of the products. The cells are usually the key component of tissue-engineered tissue, providing the tissue specificity and bioactivity required to achieve a therapeutic effect. For the cryopreservation of cultured epithelial and mucosal cell sheets, glycerol has been reported to be superior to DMSO [6]. On the other hand, DMSO is commonly used as a cryoprotectant for corneal grafts, which have been successfully applied in many clinical cases [7]. However, there has been no previous report concerning the effects of cryopreservation on cultured corneal epithelial cells. In the present study, we focused on the influence of these routinely used cryoprotectants and the protocols of cryopreservation for tissue-engineered corneal epithelial cell sheets.

We examined 3 variables: cryoprotectant (DMSO or glycerol), storage temperature (-80°C or -196°C), and the preservation period (4 weeks or 12 weeks of storage). After freezing storage, the effects of cryopreservation on histology and cell viability were analyzed. In what follows the cryopreservation method is presented in detail.

Cell Preparation and Culture
Male Japanese white rabbits were used in this study. Normal corneal limbal tissue was obtained from the unilateral corneal limbus under topical anesthesia. The biopsy sample was rinsed twice in a PBS solution containing antibiotics for 30 minutes at 37°C. It was then treated with 0.25% trypsin solution for 30 minutes at room temperature for cell dissociation. The enzyme activity was neutralized by washing with DMEM containing 10% FCS. The specimen was stirred in DMEM containing 5% FCS for 30 minutes. The cell suspension was filtered through nylon gauze (50 mm), and a suspension of purified corneal cells was obtained. The cell suspension was centrifuged for 5 minutes at 1500 rpm, and the cell pellet was resuspended in culture medium. 3T3-J2 cells were kindly provided by Dr Howard Green, and the cells were treated with 4 mg/ml of mitomycin C (MMC) in DMEM without FCS for 2 hours before epithelial cell inoculation. The 3T3-J2 cells were rinsed with PBS twice to remove MMC. A 3:1 mixture of DMEM and Ham F12 medium was supplemented with the following: FCS (5%), insulin (5 mg/ml), transferrin (5 mg/ml), hydrocortisone (0.4 mg/ml), cholera toxin (10 ng/ml), triiodothyronine (2 nmol), penicillin (100 U/ml), kanamycin (0.1 mg/ml), and amphotericin (0.25 mg/ml). Human recombinant epidermal growth factor was added at 10 ng/ml when cell adhesion was complete. The 3T3-J2 cells treated with MMC were inoculated into a cell culture dish (10 ml) at a density of 2.0×10^4 cells/cm^2. A collagen membrane filter 33 mm in diameter was placed on the feeder layer, and the epithelial cells were inoculated on the collagen membrane at a density of 1.0×10^4 cells/cm^2. The culture dish was kept in an atmosphere of 10% carbon dioxide. The medium was changed every 2 days. Stratified corneal epithelial cell sheets were obtained after 20 days (Fig. 7).

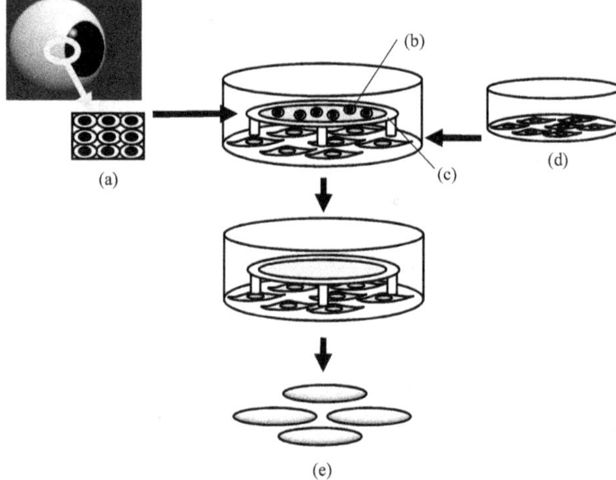

Fig. 7. Preparation for the cultured corneal epithelial cell sheets on the collagen filter membrane. (a) Limbal tissue from corneal limbus. (b) Isolated corneal epithelial cells. (c) Collagen filter membrane. (d) 3T3 cells treated with mitomycin C. (e) Culture corneal epithelial sheets (From Kito et al. 2005. Reprinted with permission).

Preparation of samples for cryopreservation
The cell sheets were punched out with a biopsy punch (8 mm in diameter) and mechanically removed from the sheets on the collagen membrane to avoid the effect of enzymes. In total, 68 samples were used for the viability analysis, and 9 samples were used for the histologic analysis.

Freezing protocols for corneal epithelial cell sheets
The freezing medium was prepared as 10% cryoprotectant and 15% FCS added to DMEM. Glycerol and DMSO were used as cryoprotectants. Cultured corneal epithelial cell sheets were frozen under a slow freezing schedule in a 1.8-mL cryotube with 1.5 ml of a freezing medium. The sheets with glycerol as a cryoprotectant were equilibrated for 40 minutes at 4°C, whereas those with DMSO were equilibrated for 20 minutes at 4°C as indicated in a previous report [8]. The incubation period was provided to allow for equilibration of cryoprotectants within tissues. The samples were incubated longer in glycerol because it penetrates the cell membrane more slowly than DMSO [9]. The 4°C incubation temperature for DMSO was selected to minimize the potential toxic effect [10]. The cultured corneal epithelial cell sheets were divided into 8 groups by storage temperature, cryoprotectant used, and storage period, i.e., samples cyropreserved with glycerol at -80°C for 4 weeks (n = 7) or 12 weeks (n = 9), samples cryopreserved with glycerol at -196°C for 4 weeks (n = 7) or 12 weeks (n = 7), samples cryopreserved with DMSO at -80°C for either 4 weeks (n = 7) or 12 weeks (n = 11), and samples cryopreserved with DMSO at -196°C for 4 weeks (n = 10) or 12 weeks (n = 10). The samples were cooled in a controlled rate freezing chamber placed in a -80°C freezer at a rate of 21°C/min from +4 to -80°C after equilibration with the cryoprotectants. Then, the samples stored at -196°C were transferred directly to the liquid nitrogen tank after overnight incubation at 280°C (Fig. 8).

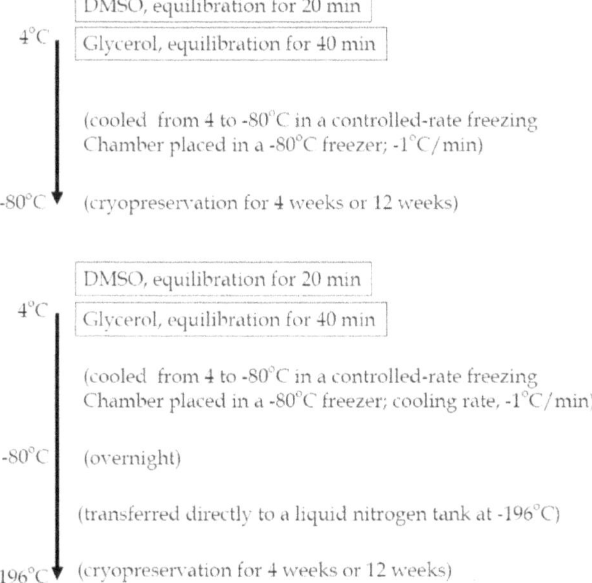

Fig. 8. Freezing protocols. Cultured corneal epithelial cell sheets were frozen under a slow freezing schedule. The sheets with glycerol as a cryoprotectant were equilibrated for 40 minutes at 4°C, whereas those with DMSO were equilibrated for 20 minutes at 4°C. An incubation period was provided to allow for equilibration of cryoprotectants within tissues. The samples were incubated longer in glycerol because it penetrates the cell membrane more slowly than DMSO. The 4°C incubation temperature for DMSO was selected to minimize potential toxic effects (From Kito et al. 2005. Reprinted with permission).

Thawing protocols
After cryopreservation, the samples were placed in a water bath at 37°C and thawed rapidly under continuous gentle stirring. The warming rate could not be measured in this study; however, the average rate under a similar condition was reported at approximately 150°C/min during the entire thawing process [8]. When the frozen cryoprotectant solution had nearly all melted, the cryotubes were removed from the water bath, and the samples were transferred into culture dishes. To avoid cell damage from osmotic stress, 6 ml of chilled DMEM containing 10% FCS was gradually added, and the melted cryoprotectant solution was diluted to less than 2% of DMSO or glycerol concentration. After equilibration for several minutes, the culture dishes were emptied of all but about 1.5 ml of diluted cryoprotectant solution. Then a second gradual dilution was performed to achieve a cryoprotectant concentration of less than 0.5%. All of the diluted cryoprotectant solution was then decanted and replaced with new culture medium. The cell sheets were allowed to equilibrate for several more minutes and washed again with the same medium.

Structural damage such as vacuolar degeneration was more clearly observed in the corneal epithelial cell sheets cryopreserved with DMSO than those with glycerol, especially at -80°C, whereas only minor morphologic changes were observed in the corneal epithelial cell sheets cryopreserved in glycerol at both temperatures. Colorimetric cell viability assay revealed that the storage conditions at the lower temperature (-196°C) showed higher cell survival than those at the higher storage temperature (-80°C). The difference between the two cryoprotectants, however, was not significant. Among the conditions used in this study, the samples cryopreserved with glycerol at -196°C showed the highest cell survival rate (70.3 ± 8.3% and 66.4 ± 14.7% for 4 and 12 weeks, respectively). The difference between this group and those stored at -80°C was significant for both 4 and 12 weeks of storage using either glycerol or DMSO. Although the cryopreserved cell sheets could not maintain their original layered structure after thawing, viable cell sheets could be regenerated.

The cell survival rates obtained after freezing storage were reasonable; however, the cryopreserved sheets could not maintain their original layered structure. More works needs to be done to better preserve the corneal epithelial cell sheets.

References

1. Rheinwald JG, Green H. Serial cultivation of strains of human epidermal keratinocytes: the formation of keratinizing colonies from single cells. Cell. 6: 331,1975
2. Pellegrini G, Traverso CE, Franzi AT. Long-term restoration of damaged corneal surfaces with autologous cultivated corneal epithelium. Lancet. 349: 990,1997
3. Dua HS, Azuara-Blanco A. Limbal stem cell of the corneal epithelium. Surv Ophthalmol. 44: 415,2000
4. Green H, Kehinde O, Thomas J. Growth of cultured human epidermal cells into multiple epithelia suitable for grafting. Proc Natl Acad Sci USA. 76: 5665,1979
5. Polge C, Smith AU, Parkes AS. Revival of spermatozoa after vitrification and dehydration at low temperatures. Nature. 164: 666,1949
6. Hibino Y, Hata K, Horie K. Structural changes and cell viability of cultured epithelium after freezing storage. J Cranio Maxillofac Surg. 24: 346,1996
7. Taylar MJ, Hunt CJ. Tolerance of corneas to multimolar dimethyl sulfoxide at 0°C; implications for cryopreservation. Invest Ophthalmol Vis Sci. 30: 400,1989

8. Bravo D, Rigley TH, Gibran N. Effect of storage and preservation methods on viability in transplantable human skin allografts. Burns. 26: 367,2000
9. Bank HL, Brokbank KGM. Basic Principles of Cryobiology, J Card Surg. 1: 137,1987
10. Schachar NS, McGann LE. Investigations of low-temperature storage of articular cartilage for transplantation. Clin Orthop. 208: 146,1985

(Kito K, Kagami H, Kobayashi C, Terasaki H, Ueda M)

Chapter 3

Bone

The use of dental implants in oral rehabilitation is becoming a standard method of care in dentistry. In the case of insufficient bone volume, a procedure for augmentation is needed. The ability to augment the alveolar ridge has gradually expanded the scope of implant dentistry. During the past 10 years, alveolar augmentation techniques have become established treatment modalities. Dahlin et al. reported an experimental study on rabbits involving the formation of new bone around titanium implants using the membrane technique [1]. In addition, various bone-grafting materials have been used for augmentation, including autologous grafts, freeze-dried bone grafts, hydroxyapatite and xenografts [2, 3]. Although the results of these investigations indicate that augmentation is clinically successful for various graft materials, it is questionable whether these materials, except for autologous bone, have adequate osteogenic potential and biomechanical properties [4, 5]. On the other hand, autologous bone, which currently remains the material of choice, is available for bone reconstructive procedures [6]. However, its use is limited due to donor site morbidity and limited amounts of graft material available for harvesting.

To avoid these problems, we attempted to regenerate bone in a significant osseous defect with minimal invasiveness, and to provide a clinical alternative to the graft materials described above. The new technology that we developed is called "injectable tissue-engineered bone", and involves the morphogenesis of new tissue using constructs formed from isolated cells with biocompatible scaffolds and growth factor, and was established based on tissue engineering concepts [7-9].

Mesenchymal stem cells

Mesenchymal stem cells (MSCs) are frequently used for bone tissue engineering and increasingly applied in the clinic. Although engineering bone tissue using MSCs is feasible, the size of the regenerated bone is limited by nutrient transport. The grafted cells require an oxygen/nutrition supply to survive and early neovascularization is considered essential for successful bone tissue engineering. Development of an efficient neovascularization method to sustain the engineered implants is clinically important. The relationship between vascular endothelial factors (e.g. VEGF) and bone regeneration emphasizes the important role of vasculature not only for survival but also for the proper formation of tissue-engineered bone.

In 1997, Asahara et al. characterized endothelial progenitor cells (EPCs) in human peripheral blood using magnetic beads selection [10]. Since EPCs can give rise to endothelial cells (ECs) and are known to facilitate the neovascularization of an implanted site, EPCs may be used to facilitate collateral vessel growth into ischemic tissues through delivery of antiangiogenic or proangiogenic agents. EPCs have been implanted into various ischemic tissue models, for example, ischemic hindlimbs and areas of myocardial infarction. Recently, EPCs have also been used to engineer blood vessels. Taken together, it is conceivable that supplementing osteogenic cells with ECs or EPCs may facilitate osteogenesis in tissue-engineered bone constructs. However, little is known about how EPCs may influence development of tissue-engineered bone.

We investigated the potential of EPCs to facilitate neovascularization in implants and evaluated their influence on bone regeneration. The influence of EPC soluble factors on osteogenic differentiation of MSCs was tested by adding EPC culture supernatant to MSC culture medium. To evaluate the influence of EPCs on MSC osteogenesis, canine MSCs-derived osteogenic cells and EPCs were seeded independently onto collagen fiber mesh scaffolds and cotransplanted to nude mice subcutaneously. Results from co-implant experiments were compared to implanted cells absent of EPCs 12 weeks after implantation. Factors from the culture supernatant of EPCs did not influence MSC differentiation. Co-implanted EPCs increased neovascularization and the capillary score was 1.6-fold higher as compared to the MSC only group ($P<0.05$). Bone area was also greater in the MSC + EPC group ($P<0.05$) and the bone thickness was 1.3-fold greater in the MSC + EPC group than the MSC only group ($P<0.05$). These results suggest that soluble factors generated by EPCs may not facilitate the osteogenic differentiation of MSCs; however, newly formed vasculature may enhance regeneration of tissue-engineered bone (Fig. 9).

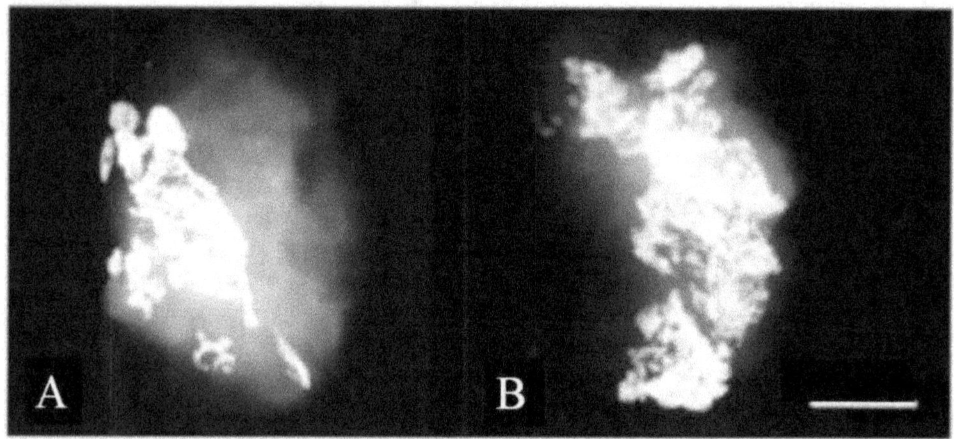

Fig. 9. Soft X-ray radiographs of implants 12 weeks after implantation. A: MSC only group: MSC-derived osteogenic cells were seeded onto collagen, which was then wrapped with the same scaffold without cells. B: MSC and EPC group: MSC-derived osteogenic cells were seeded onto one scaffold, then covered with another scaffold seeded with EPCs (From Usami et al. 2006. Reprinted with permission).

Tissue engineered bone 'injectable bone'

We previously reported that tissue-engineered bone induces excellent bone regeneration and promotes bone formation in a grafted area treated with platelet-rich plasma (PRP), which contains various growth factors [8].

After a period of housing, 12 adult hybrid dogs with a mean age of 2 years were operated dMSCs were isolated from 10 ml samples of dog iliac bone marrow aspirates. Bone marrow cell isolation and expansion was performed according to previously published methods [11]. Briefly, basal medium (condition medium), low-glucose Dulbecco's modified Eagles medium (DMEM) and growth supplements consisting of 50 ml of mesenchymal cell growth supplement, 10 ml of 200 mM L-glutamine and 0.5 ml of a penicillin–streptomycin mixture containing 25 U of penicillin and 25 µg of streptomycin. The three supplements used for inducing osteogenesis, dexamethasone (Dex), sodium β-glycerophosphate (β-GP) and L-ascorbic acid 2-phosphate (AsAP). The cells were incubated at 37°C in a humidified atmosphere containing 95% air and 5% CO_2. dMSCs were replated at a density of 3.1×10^3 cells/cm^3 in 0.2 ml/cm^2 condition medium. The implants were assessed by histological and histomorphometric analysis, 2, 4 and 8 weeks after implantation. The implants exhibited varying degrees of bone-implant contact (BIC). The BIC was 17%, 19% and 29% (control), 20%, 22% and 25% (fibrin), 22%, 32% and 42% (dMSCs/fibrin) and 25%, 49% and 53% (dMSCs/PRP/fibrin) after 2, 4 and 8 weeks, respectively. This study suggests that tissue-engineered bone may be of sufficient quality for predictable enhancement of bone regeneration around dental implants when used simultaneous by with implant placement. Recent tissue engineering approaches have attempted to create new bone based on the use of MSCs seeded onto porous ceramic scaffolds with osteoconductive properties [12] (Fig. 10). These attempts have yielded sub-optimal results due to the slow resorption rate of hydroxyapatite-based ceramics. Also, these delivery substances do not exhibit good plasticity and the cellular implantation procedure is complicated by problems associated with the delivery vehicles because the block materials do not have plasticity. Isogai et al. reported that a combination of fibrin glue with delivery vehicle and cultured periosteal cells resulted in new bone formation at heterotopic sites in nude mice [13]. In numerous reports about materials, fibrin was found to have hemostatic effects and to promote wound healing. In a bone regeneration study using the rabbit tibia, Bosch et al. and Schwarz et al. reported that fibrin stimulated neovascularization of bone with accelerated healing and earlier new bone formation [14, 15]. Additionally, the use of fibrin as an osteoconductive material has been recommended [16]. Therefore, we used fibrin as a scaffold, which is one of the three key factors in the tissue engineering concept [9].

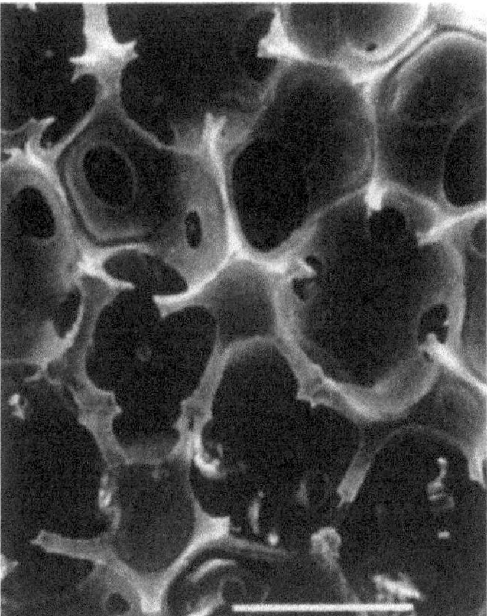

Fig. 10. Scanning electron microscopy photomicrograph of a cross-section of β-tricalcium phosphate. Bar = 500 µm (From Boo et al. 2002. Reprinted with permission).

Distraction osteogenesis using tissue engineered bone

Distraction osteogenesis has become a widely accepted technique for reconstructing bone defects in the maxillofacial region. This technique provides autologous and predictable bone formation without grafting procedures but requires long-term treatment that includes latent, lengthening, and consolidation periods. The long treatment time results in a high rate of complications such as infection, pin loosening, and fracture. The recommended rate of gradual bone lengthening as described by Ilizarov is 1 mm per day [17]. A lower rate of distraction tends to result in bony union, whereas a higher rate of distraction may delay bone union or result in fibrous union.

To promote bone formation and shorten the consolidation period, some attempts at applying hyperbaric oxygenation or electrical, ultrasonic, or chemical stimulation have been made [18]. Several recent studies have shown that injecting cells with osteogenic potential into distracted callus enhances its consolidation, but there have been few attempts at higher-rate distraction [19-22].

We previously reported on a tissue-engineered osteogenic material (TEOM) [23]. This material is an injectable gel of autologous MSCs, which are culture-expanded then induced to be osteogenic in character, and PRP activated with thrombin and calcium chloride. The injection of TEOM into the distraction gap has advantages. MSCs can be expanded and induced to osteoblastic lineage *ex vivo*. Moreover, both MSCs and PRP are autologous materials.

Bilateral maxillary distraction was performed at a higher rate in rabbits to determine whether locally applied TEOM enhances bone regeneration (Fig. 11). The material was an injectable gel composed of autologous mesenchymal stem cells, which were cultured then induced to be osteogenic in character, and PRP. After a 5-day latency period, distraction devices were activated at a rate of 2.0 mm once daily for 4 days. Twelve rabbits were divided into 2 groups. At the end of distraction, the experimental group of rabbits received an injection of TEOM into the distracted tissue on one side, whereas, saline solution was injected into the distracted tissue on the contralateral side as the internal control. An additional control group received an injection of PRP or saline solution into the distracted tissue in the same way as the experimental group. The distraction regenerates were assessed by radiological and histomorphometric analyses. The radiodensity of the distraction gap injected with TEOM was significantly higher than that injected with PRP or saline solution at 2, 3, and 4 weeks postdistraction. The histomorphometric analysis also showed that both new bone zone and bony content in the distraction gap injected with TEOM were significantly increased when compared with PRP or saline solution.

TEOM injection at the end of distraction promotes new bone formation following a higher rate of distraction. TEOM injection may be able to compensate for the insufficient distraction gap at a higher rate (Fig. 12).

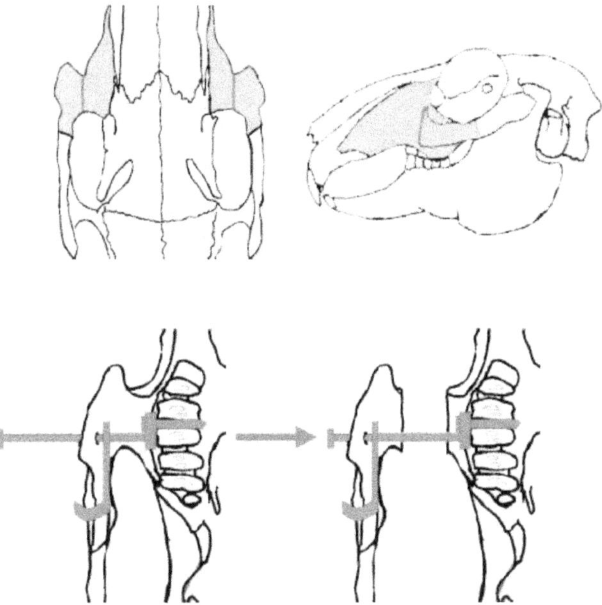

Fig. 11. A: Schematic drawing of the maxilla and maxillary distraction. Red line, osteotomized line; yellow area, maxilla.

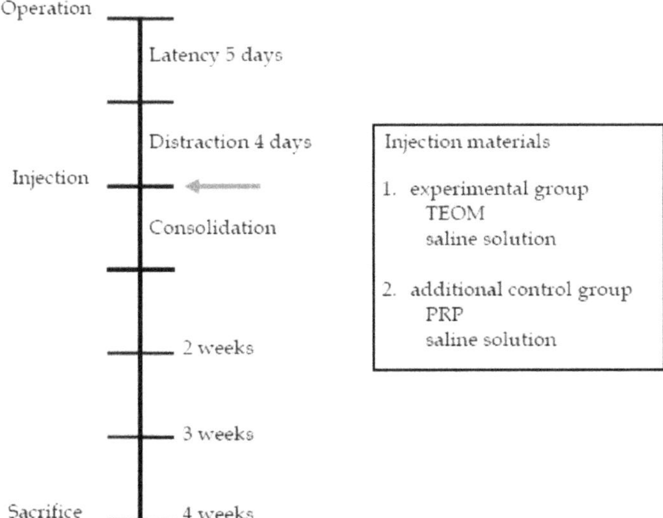

Fig. 11. B: Distraction protocol and experimental design (From Kinoshita et al. 2008. Reprinted with permission).

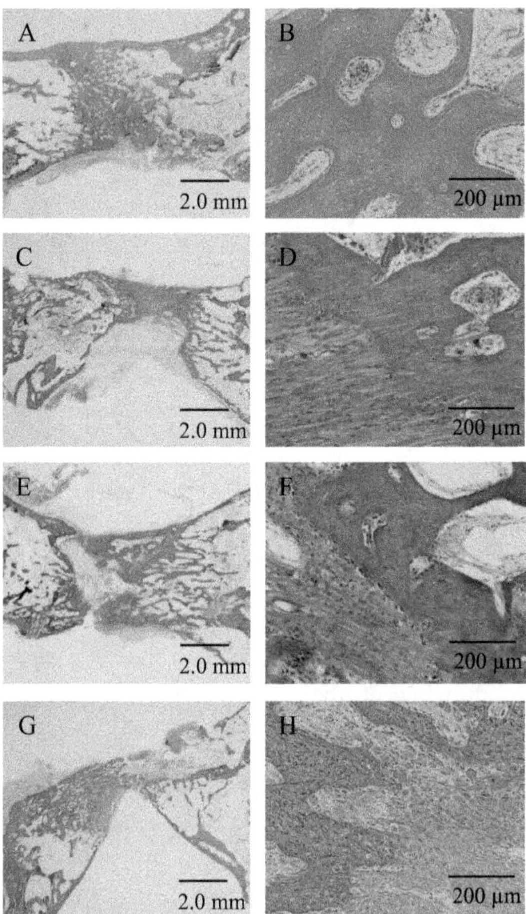

Fig. 12. Histological view of the distracted maxilla, staining H-E. Experimental group A: TEOM-injected side; B: center of the distraction zone on the TEOM-injected side; C: saline solution-injected side; D: center of the distraction zone on the saline solution-injected side. Additional control group, E: PRP-injected side; F: center of the distraction zone on the PRP-injected side; G: saline solution-injected side; H: center of the distraction zone on the saline solution-injected side (From Kinoshita et al. 2008. Reprinted with permission).

Bone regeneration using 'periosteum'

The periosteum is comprised of two tissue layers: the outer fibroblast layer that provides attachment to soft tissue, and the inner cambial region that contains a pool of undifferentiated mesenchymal cells, which support bone formation [24]. Recently, studies have reported the existence of osteogenic progenitors, similar to MSCs, in the periosteum [25, 26]. Under the appropriate culture conditions, periosteal cells secrete extracellular matrix and form a membranous structure [27]. The periosteum can be easily harvested from the patient's own oral cavity, where the resulting donor site wound is invisible. Owing to the above reasons, the periosteum offers a rich cell source for bone tissue engineering.

Our group has previously demonstrated bone regeneration using a cultured periosteum (CP) in a critical-sized rat calvaria bone defect [28] (Fig. 13). CP has also been shown to regenerate bone in a surgically created furcation bone defect using a canine model [27]. Considering the biocompatibility of CP and its capacity for alveolar bone regeneration, it should be useful to investigate the potential of CP for bone regeneration around an implant site. The purpose of this study was to investigate the potential of CP to regenerate bone to mitigate implant dehiscence defects.

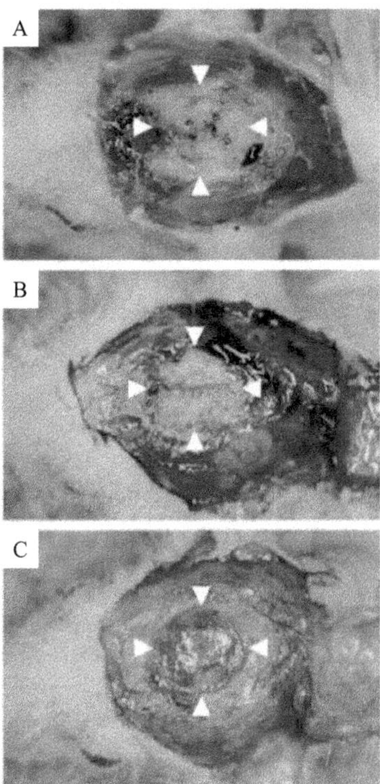

Fig. 13. A–C: Photographs showing representative calvarial bone defect of athymic rats 3 months after surgery. A: Animals with grafted fresh CP showed complete closure of calvarial defect. B: Animals, grafted with cryopreserved CP also showed complete closure of the defect, and there was no apparent difference between fresh and cryopreserved CP macroscopically. C: However, the bone defect of the control group without CP remained almost the same size as before grafting. Arrowheads indicate margin of original bone defect (From Mase et al. 2006. Reprinted with permission).

Four healthy beagle dogs were used in this study. Implant dehiscence defects (4 × 4 × 3 mm) were surgically created at mandibular premolar sites where premolars had been extracted 3 months back. Dental implants (3.75 mm in diameter and 7 mm in length) with machined surfaces were placed into the defect sites (14 implants in total). Each dehiscence defective implant was randomly assigned to one of the following two groups: (1) PRP gel without cells (control) or (2) a periosteum cultured on PRP gel (experimental). Dogs were killed 12 weeks after operation and nondecalcified histological sections were made for histomorphometric analyses including percent linear bone fill (LF) and bone-implant contact (BIC) (Fig. 14). Bone regeneration in the treatment group with a CP was significantly greater than that in the control group and was confirmed by LF analysis. LF values in the experimental and the control groups were 72.36 ± 3.14% and 37.03 ± 4.63%, respectively ($P<0.05$). The BIC values in both groups were not significantly different from each other. The BIC values in the experimental and the control groups were 40.76 ± 10.30% and 30.58 ± 9.69%, respectively ($P = 0.25$) and were similar to native bone.

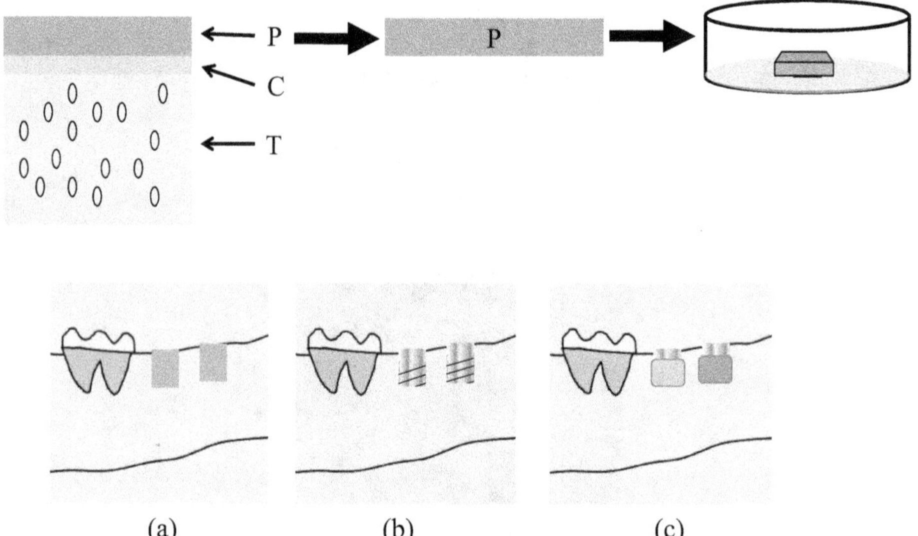

Fig. 14. Schematic drawing of the surgical procedures used in this study (P, Periosteum; C, Cortical bone; T, Trabecular bone). (a) Dog mandibular defect model (4 × 4 × 3 mm). (b) Implant placement. (c) Transplantation PRP gel (left) and cultured periosteum membrane on PRP gel (right) (From Mizuno et al. 2008).

The consensus tissue engineering paradigm includes cells, scaffolds, and bioactive molecules. For periodontal therapy, there are several reports based on this tissue engineering paradigm that incorporate various polymers such as collagen and gelatin as a scaffold material [29-31]. However, natural biodegradable materials cannot eliminate the possible risk of infection and degraded products may interfere with the regeneration process. Instead of culturing cells on natural biodegradable scaffolds, we have been able to stimulate periosteal cells to form their own matrix and generate a cell-populated membrane *in vitro* [27]. This method creates a CP durable enough to be held by forceps, making it feasible to transplant CP without the need for a biodegradable support. Furthermore, the thickness of the CP is approximately 200 μm, which may be beneficial to cells, allowing oxygen and nutrients to diffuse into the transplanted tissue (Fig. 15).

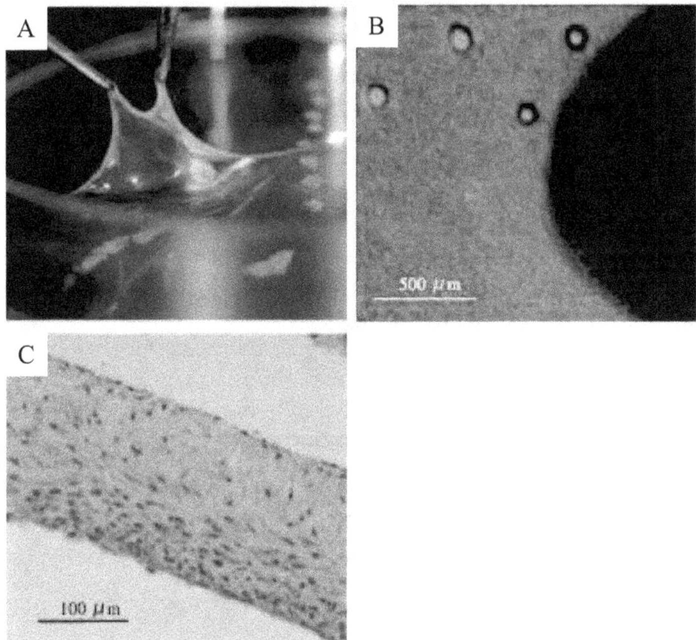

Fig. 15. A, B: CP macroscopic and microscopic findings. Bovine periosteal cells were cultured for 4 weeks. A: The cells became a membranous structure with enough mechanical strength to handle with forceps. Cells can be obtained by a conventional explant culture of periosteal fragment. B: Phase-contrast photomicrograph showing the cultured CP. C: Photomicrograph showing hematoxylin and eosin staining of CP section. The CP is approximately 100–300 µm in thickness and consists of 20–30 cellular layers (From Mase et al. 2006. Reprinted with permission).

A major disadvantage of using CP for clinical treatment might be the time period required for tissue culture. Four to 6 weeks is a typical time frame for obtaining CP with enough mechanical strength to be transplanted. Furthermore, the cultured period would likely differ for each patient and may be unpredictable at the beginning of culture. This uncertainty makes it difficult to formulate a treatment schedule in advance. To overcome this problem, we have investigated the potential of CP cryopreservation [28]. In this study, the optimal preincubation protocol for CP was investigated and it was found that CP could be successfully cryopreserved under specific conditions without loss of osteogenic potential. Cryopreservation of CP should increase the usefulness of these approaches in future clinical applications. This study demonstrated the feasibility of a CP to regenerate bone at implant dehiscence defect.

Translational Research

Translational research involves application of basic scientific discoveries into clinically germane findings and, simultaneously, the generation of scientific questions based on clinical observations. At first, as basic research we investigated tissue-engineered bone regeneration using MSCs and PRP in a dog mandible model. We also confirmed the correlation between osseointegration in dental implants and the injectable bone. Bone defects made with a trephine bar were implanted with graft materials as follows: PRP, dog MSCs (dMSCs) and PRP, autologous particulate cancellous bone and marrow (PCBM), and control (defect only). Two months later, dental implants were installed. According to the histological and histomorphometric observations at 2 months after implants, the amount of BIC at the bone-implant interface was significantly different between the PRP, PCBM, dMSCs/PRP, native bone, and control groups. Significant differences were also found between the dMSCs/PRP, native bone, and control groups in bone density. These findings indicate that the use of a mixture of dMSCs/PRP will provide good results in implant treatment compared with that achieved by autologous PCBM. We then applied this injectable tissue-engineered bone to onlay plasty in the posterior maxilla or mandible in three human patients. Injectable tissue-engineered bone was grafted and, simultaneously, 2-3 threaded titanium implants were inserted into the defect area. The results of this investigation indicated that injectable tissue-engineered bone used for the plasty area with simultaneous implant placement provided stable and predictable results in terms of implant success. We regenerated bone with minimal invasiveness and good plasticity, which could provide a clinical alternative to autologous bone grafts. This might be a good case of translational research from basic research to clinical application.

Clinical application

From these animal experiments, we adopted "injectable tissue-engineered bone" to regenerate bone in a significant osseous defect that was minimally invasive and had good plasticity, and to provide a clinical alternative to the graft materials mentioned above. One of the advantages of injectable tissue-engineered bone is the use of autologous cells for bone regeneration. Tissue engineering technology by autologous cell transplantation is one of the most promising therapeutic concepts being developed because it may solve problems, including donor site morbidity from autologous grafts, immunogenicity of allogenic grafts, and loosening of alloplastic implants. In this method, we use differentiated bone marrow derived stem cells (BMDSCs) as isolated cells for bone formation, and PRP as a growth factor and scaffold. In our hospital, we experienced many clinical cases using this method, such as maxillary sinus augmentation, periodontal treatment, and distraction [32-34]. We here report the procedures and results for these cases.

Preparation of cells
One and a half month before surgery, BMDSCs were isolated from the patient's iliac crest bone marrow aspirates (20 ml) and cultured. The control medium contained the following: basal medium, low-glucose DMEM, 10% patient serum or 10% fetal bovine serum, and growth supplements (10 ml of 200 mM L-glutamine and 0.5 ml of a penicillin-streptomycin mixture containing 25 units of penicillin and 25 µg of streptomycin). Each patient could choose the type of serum (patient serum or fetal bovine serum) for cultivation of BMDSCs. For human serum preparation, human blood was isolated in a 200 ml collection bag under sterile conditions. Subsequently, the blood collected was centrifuged at 3500 rpm for 10 minutes and the supernatant was collected. Three supplements for inducing osteogenesis—100 nM of dexamethasone, 10 mM of sodium β-GP, and 80 µg/ml of AsAP. Cells were incubated at 37°C in a humid atmosphere containing 95% air and 5% carbon

dioxide. Differentiated BMDSCs were trypsinized and used for implant placement. To verify the safety of cultured cells, the culture media were examined for contamination by bacteria, fungi, or mycoplasmas before transplantation.

Preparation of PRP
Preoperative hematology included complete blood count with platelet counts. PRP, extracted 1 day before surgery, was isolated in a 200 ml collection bag containing an anticoagulant, citrate, under sterile conditions in the Blood Transfusion Service of Nagoya University Hospital. Briefly, the blood collected was first centrifuged at 1500 rpm for 10 minutes. Subsequently, the yellow plasma (containing the buffer coat that contained platelets and leukocytes) was removed. The second centrifugation was conducted at 3500 rpm for 5 minutes to combine platelets with a single pellet. The plasma supernatant, which was platelet-poor plasma and contained a relatively small number of cells, was removed. The resulting pellet of platelets, the buffer coat/plasma fraction (PRP), was resuspended in remaining 20 ml of the plasma before use in the platelet gel.

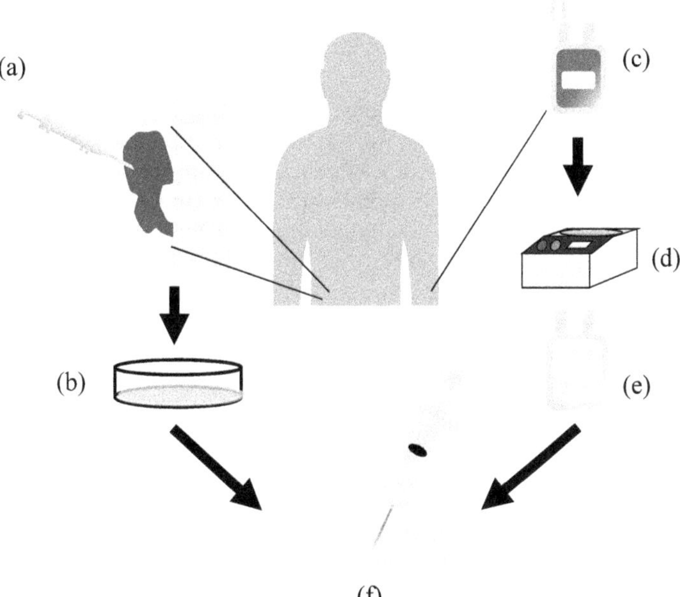

Fig. 16. Protocol of tissue engineered bone. (a) Harvest of bone marrow. (b) Cell culture and osteoinduction. (c) Collection of whole blood. (d) Centrifuge. (e) Platelet rich plasma. (f) Injection of tissue engineered bone (From Ueda et al. 2008).

Preparation of injectable tissue-engineered bone
PRP was stirred and stored at 22°C in a conventional shaker until used. Powdered human thrombin (500 units) was dissolved in 10% calcium chloride in a separate sterile cup. The PRP, BMDSCs (5.0×10^6 cells/ml), and air were aspirated into the first 2.5 ml sterile syringe. The thrombin-calcium chloride mixture (300 μl) was aspirated into the second 2.5 ml syringe. The two syringes were connected with a T connector, and the plungers of the syringe were

pushed and pulled alternatively, allowing air bubbles to flow back and forth between the two syringes. Within 5-30 seconds, the contents gained gel-like consistency because thrombin affected the polymerization of fibrin to produce an insoluble gel (Fig. 16).

Application for maxillary sinus augmentation

After tooth loss, alveolar augmentation of the extensively atrophied maxillary process may be required to restore the masticatory function of the patient by means of substitute teeth anchored on dental implants. In order to obtain adequate volume of bone enough to insert dental implants, elevation of the maxillary sinus floor has been carried out as a routine clinical procedure for more than 15 years [35-39]. Where the bone thickness between the maxillary sinus and the alveolar crest is less than 8 mm, sinus floor elevation without bone graft materials is insufficient. A bone graft-induced increase in the thickness of the alveolar sinus floor is necessary to support longer implants that are required [40, 41]. The success of the dental implants is to be evaluated over a long period of time.

Sixteen sinus augmentations in 12 patients, partially or totally edentulous patients 44-60 years of age (mean age: 54 years), were performed. All the patients had a conventional problem of denture retention due to severe posterior alveolar ridge atrophy; the average height of their residual sinus floor was <6 mm, for which sinus graft and dental implants would solve the problem.

After routine oral and physical examinations, patients who did not desire to undergo any surgery for harvesting autologous bone were selected for injectable tissue-engineered bone grafting. They were healthy and free of any disease that might affect treatment outcomes (e.g., diabetes, immunosuppressive chemotherapy, and rheumatoid arthritis). Each patient was given detailed information about the intervention, including surgical techniques, types of graft material and dental implants, and the uncertainties of conducting a new bone-regenerative procedure. Informed consent in writing was requested of each patient.

Surgical techniques

Sinus augmentation was conducted under general anesthesia. The sinus grafting procedure followed Tatum's classical description [35]. Briefly, the mucoperiosteal flap was elevated to create a trap door with a round hollow burr in the lateral wall of the maxillary sinus. After mobilization, the door was reflected inward. The space created by this procedure was filled with 1.8-5.4 g of injectable tissue-engineered bone to simultaneously place dental implants. The mucoperiosteal flap was repositioned and sutured in the usual manner. After surgery, patents received cephalosporins (300 mg/day) as antibiotics, and loxoprofen sodium (180 mg/day) as analgesics for 3 days.

Postoperative course

The incidences of grafted bone resorption and implant loss after sinus augmentation with various bone substitutes have been recorded. The complete resorption of bone substitutes, especially autologous bone, was observed in 2.7% of patients [37]. However, this regenerated bone did not show resorption and remained in the sinus floor that had been elevated by injectable tissue-engineered bone. The mineralized tissue 2 years after operation increased by 8.8 ± 1.6 mm, presumably due to the properties of tissue-engineered bone (Fig. 17).

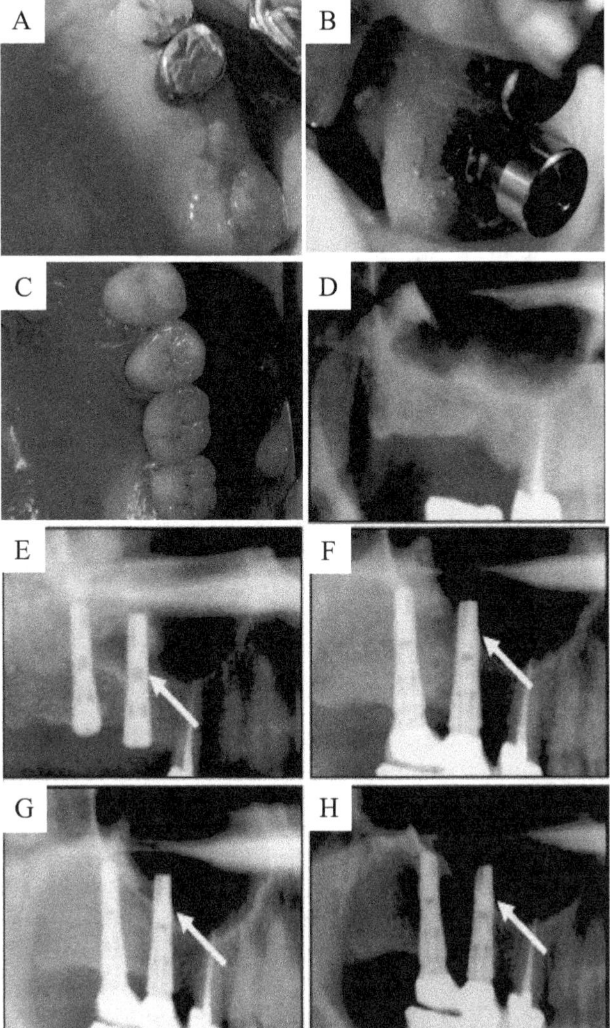

Fig. 17. A: Preoperative macro view. B: Observation of second-stage surgery 6 months after the implant installation. The exposed thread was surrounded by newly formed bone and confirmed successful osseointegration. C: Last prosthesis observation by porcelain fused to a metal crown. These did not exceed 2 mm, and a healthy and firm peri-implant mucosa had been established. D: Panoramic radiograph, preoperative. E: Panoramic radiograph, postoperative. F: Panoramic radiograph, postoperative 1 year. G: Panoramic radiograph, postoperative 2 years. H: Panoramic radiograph, postoperative 3 years (From Ueda et al. 2008. Reprinted with permission).

For long periods of time, maxillary sinus floor augmentation has constituted a surgical procedure to gain bone mass required for placing dental implants. Further, there is also consensus that some threshold of osseous deficiency, vertical, horizontal, or both, exists at a site where a sinus bone graft is required for successful implant treatment regardless of residual bone quality. If there is vertical bone less than 8 mm in height in the posterior maxilla, sinus floor elevation without bone graft materials is insufficient. A sinus graft should be strongly recommended to provide adequate support for the placement of dental implants [40, 41]. In these cases, the average residual bone height was 5.5 ± 1.6 mm (range: 2-10 mm); therefore, we applied tissue-engineered bone as a graft material for the sinus graft. On the other hand, in our previous study [8, 23, 42-44], we found that the tissue-engineered bone was well-formed mature bone, and the bone-regenerating ability increased significantly compared to the nongrafted control of defect-only sites or PRP-only sites that had been reported not to function effectively for bone regeneration [45], confirming the radiological and histological data [8, 45]. In addition, we measured the values from a Vickers hardness test, which indicates mechanical properties for bone formation. The values of nongrafted control, PRP, autologous bone, and tissue-engineered bone were 8, 9, 13, and 17, respectively, 2 weeks after operation [44]. Moreover, in our experiment we used rabbit maxillary sinus that has well-defined ostium similar to that of humans. The augmented height and bone volume showed peaks as early as 2 weeks in tissue-engineered bone sites; on the other hand, the volume of newly formed bone reached a peak value within 4 weeks in autologous bone sites at 2, 4, and 8 weeks of experimental time [42, 46]. Thus bone regeneration may indicate early bone formation and enhanced bone quality as the main advantages of BMDSCs, since our results were consistent with the report of our previous animal experiments [8, 23, 44].

This technique might be effective for maxillary sinus augmentation. This result also might be the effect of this grafted material, tissue-engineered bone by BMDSCs and PRP, and not the effect of sinus membrane elevation alone. The BMDSCs in the bone marrow are induced in cells with osteogenic capacity, and the MSCs in BMDSCs are considered more feasible for tissue engineering because the former proliferates faster due to a lower degree of differentiation. The PRP contains not only fibrinogen, which forms a fibrin network acting as a matrix, but also cytokinetic substances such as PDGF, TGF-β, IGF, and VEGF, which can stimulate MSCs to transform into osteoblasts [47]. The growth factors are believed to have an osseous regenerative effect on the MSCs and contribute to cellular proliferation, matrix formation, collagen synthesis, osteoid production, and other processes that accelerate tissue regeneration. However, further research will be required to examine the effect.

Additionally, osseointegration is the most important condition for success in dental implant treatment. In clinical cases, implant loss occurs with no osseointegration because of infection, absorption and loss of bone volume. In our other study, we showed that the surface of implants attached to regenerated bone by tissue-engineered bone. Further, no infection was observed, and regenerated bone volume was not reduced 2 years after operation from the above-mentioned result. As a result, no implant loss was observed whatever because our procedures used autologous graft materials, autologous BMDSCs and PRP, in our cases (Fig. 18).

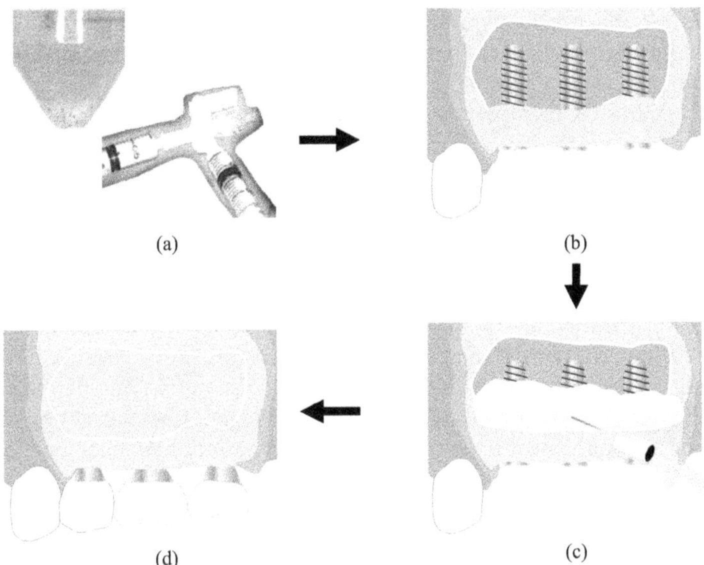

Fig. 18. Protocol of tissue engineered bone for sinus lift. (a) Mixing osteoblasts and PRP. (b) Installation of dental implants. (c) Injection of tissue engineered bone. (d) Bone regeneration (From Ueda et al. 2005).

Application for periodontal treatment

Periodontal disease is an infectious disease that affects tooth-supporting tissues. Clinically, the color changes of the gingival, periodontal pocket formation, bleeding, clinical attachment loss of alveolar bone as detected on radiolucent disease is due to bacterial plaque in the periodontal pockets. The main aim of conventional periodontal therapy is to halt and possibly reverse the attachment loss resulting from the disease. To this end, initial therapy is focused on removal of bacterial plaque from teeth and periodontal pockets and prevention of supragingival plaque accumulation. Subgingival plaque can be removed by a nonsurgical form of therapy, such as scaling and root planing, or by surgical means. The efficacy of nonsurgical methods is well documented [48].

Several studies have confirmed the efficacy of mechanical subgingival plaque control in periodontal therapy, irrespective of the approach used [49]. Adequate supragingival plaque control by patients is required for successful periodontal treatment [50]. However, subgingival bacteria in deep pockets with compared anatomy, in infrabony pockets, and in areas of furcal involvement are sometimes difficult to remove with nonsurgical therapy. In those pockets, open access with surgical therapy may be indicated to clean the root surfaces. Even when inflammation has been eliminated and healthy periodontal tissue has been established after pathogenic microorganisms are removed from periodontal pockets, the anatomy of healed defects can be a problem, particularly in areas in which esthetics is critical and maintenance is difficult, such as in dentitions with gingival recession, infrabony defects, and furcations.

In the early 1980s, a series of experimental studies was conducted on a procedure to regenerate the lost attachment apparatus. A membrane was placed under the flap to prevent epithelial downgrowth and to create space for periodontal reformation [51]. The procedure, termed guided tissue regeneration (GTR), was introduced into the clinical setting by Gottlow et al. [52]. GTR therapy has been applied to furcation and infrabony sites under certain conditions, and its efficacy has been reported [53, 54]. However, it has been claimed that the new attachment between regenerated cementum obtained by GTR procedures and root dentin may not be as strong as the attachment between the original cementum and root dentin [55]. Because cementum formed following GTR therapy is apparently different from cementum formed during tooth development (a cellular cementum), the appropriateness of the term regeneration in the context of GTR therapy has been questioned [56]. Recently, some authors have started using the term true periodontal regeneration, which has been defined as "healing after periodontal treatment that results in the regain of lost supporting tissues, including a new cellular cementum attached to the underlying dentin surface, a new periodontal ligament with functionally oriented collagen fibers inserting into the new cementum, and new alveolar bone attached to the periodontal ligament [57]." Another technique, developed recently, is the application of enamel matrix derivatives to root surfaces. However, although these treatments have been reported to be effective for periodontal tissue regeneration, the indications for such treatments are rather limited and the amounts of regenerated tissue are not predictable. This indicates that further theoretical and technical developments are needed in the field of periodontal regenerative therapies before such therapy can be widely used in daily practice.

In the previous cases, we used injectable tissue-engineered bone grafting to effectively regenerate bone for dental implant placement, and the result confirmed that tissue engineering can elicit as much bone regeneration as autologous bone grafts. So, we applied this method as periodontal regenerative therapy.

Surgical techniques

Immediately before surgery, the patient rinsed her mouth with 0.2% chlorhexidine solution for 90 seconds. The surgical area was anesthetized with lidocaine adrenaline 2%. Following pocket and releasing incisions, buccal and lingual full-thickness flaps were elevated and the epithelium was removed from the inside of the flaps. Granulation tissue residing in the defect area was carefully excised, and the root surface was scaled and planed. No bone recontouring was performed. Subsequently, MSCs-PRP gel was applied to the root surface and adjacent defect space. The flaps were replaced and closed with sutures. After 2 weeks, the sutures were removed. The patient was instructed to rinse three times daily with a 0.1% or 0.2% solution of chlorhexidine digluconate. Mechanical cleaning of the surgical site was not recommended during the first 4 postoperative weeks. Supportive care, including professional tooth cleaning, was performed every 2 months.

Postoperative course

By 1 year after this treatment, the pocket depth decreased from 5 mm to 1 mm. This clinical improvement should be predictable for teeth. Radiographic assessments revealed that the bone defect was indeed reduced in depth. But this may be a result of the short time allowed for a change in radiopacity; after a longer period, the postoperative progress would have become radiographically apparent. Whereas some regeneration may occur in humans following a regenerative surgical approach, complete and predictable true regeneration is still difficult to attain. Based on recent clinical results, GTR therapy appears to have the most promising prospects for regeneration, although its clinical efficacy and predictability in periodontal defects have yet to be thoroughly tested in controlled clinical trials [57]. Also, when the GTR approach is used to treat periodontal defects, the risk of membrane exposure must be considered as a major complication [58-60]. The reported prevalence of membrane exposure is in the 70% to 80% range [61]. In such cases, adequate membrane fixation and soft tissue coverage can be difficult to perform. On the other hand, it is easier to apply injectable tissue-engineered bone, which is a formed gel, than to position a membrane around a defect, and since gingival recession and unesthetic outcomes occur when membranes become exposed, it seems logical to use tissue-engineered bone rather than a membrane. Also, the tissue-engineered bone assumes a firm, gel-like consistency and may have the ability not only to immobilize to implants in place but also to provide a seal around the tooth. In a preliminary animal study, tissue-engineered bone prevented downgrowth of the epithelium equally well as the GTR method. In addition, it has been claimed that the new attachment between regenerated cementum obtained from GTR procedures and root dentin may not be as strong or continuous as the attachment between the original cementum and root dentin [55]. On the other hand, in a periodontal tissue regeneration study using this treatment in dogs, MSCs played an important role in cementification, and the structure of regenerated cementum was similar to that of natural cementum on roots versus that regenerated using the GTR method. Therefore, injectable tissue-engineered bone treatment might be more useful than the GTR method for true periodontal and interdental papilla regeneration (Fig. 19).

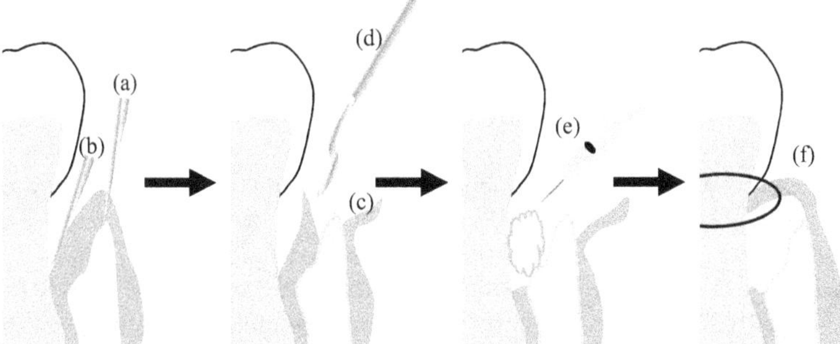

Fig. 19. Protocol of tissue engineered bone for periodontal treatment. (a) Internal bevel incision. (b) Intersulcular incision. (c) Flap elevation. (d) Scaling and root planning. (e) Injection with tissue engineered bone. (f) Suture (From Yamada et al. 2006).

Application to alveolar cleft defects

The reconstruction of alveolar cleft defects is well established, with the most widely accepted approach being secondary alveolar cleft osteoplasty in the mixed dentition phase with autologous bone grafting [62, 63]. The source material for most bone grafts has been particulate marrow harvested from the anterior iliac crest, and this represents the standard material with which other materials from rib, mandible, calvarium, and tibia are compared

[62-64]. Donor site morbidity is an important factor in deciding the site for harvesting cancellous bone. Allogenic or xenogenic materials can eliminate this concern but not the risk of disease transmission. As another solution, the use of injectable tissue-engineered bone in bone augmentation procedures as a replacement for autologous bone grafts, offers predictable results with minimal donor-site morbidity [23, 65]. We considered that tissue-engineered bone is beneficial material for alveolar cleft osteoplasty, and applied this treatment.

Surgical techniques
Following a 3-cm-long mucosal incision at the level of the labiogingival junction, dissections were made in the ingrown scar tissue to reach the bony surface of the cleft walls. The tissue was then elevated in the subperiosteal plane to the levels of the anterior nasal spine anteriorly, the lateral piriform rim superiorly and to the alveolar ridges inferiorly, while taking care not to damage the unerupted teeth and the content of the incisive canal. The flaps of the nasal floor and the oral mucosa formed the ceiling and the floor of the cleft cavity, respectively. The ceiling, floor and front walls of the defect were supported with a 0.1-mm-thick titanium-mesh plate. The thus-created pouch was filled with all the prepared TEOM through a syringe using a packer. Following release incisions in the periosteum and the scar tissue of the flaps and to allow them to cover the grafted area, the wound was closed without tension.

Postoperative course
Before this treatment, a 3-month-old female patient born with a congenital left unilateral cleft lip and alveolus underwent a cheiloplasty that resulted in no remaining oronasal fistula. At 9 years of age, computed tomograms (CTs) revealed that the left maxillary canine, lateral, and supernumerary incisors had formed half of their roots, and had closely surrounded the alveolar cleft bony defect which was 10 mm wide and 13 mm deep anteroposteriorly. The left central incisor was orthodontically overcorrected due to previous severe rotation and distal location (Fig. 20).
The patient exhibited an uneventful postoperative course. The radiopacity of serial CTs slicing the middle level of the alveolar cleft in the grafted region increased gradually over time. Dome-shaped radiopaque images with 233 Hounsfield units (HU) faced one another, extended from the cleft bony walls inside the cavity after 3 months, and were fused together into an image with 324 HU after 6 months. The image increased in radiopacity to 447 HU in 9 months, and at the bony bridge the lateral and supernumerary incisors horizontally migrated from their original positions in the respective major and minor segments. The incisive canal was reconstructed just medial to the bridge. The erupting canine and lateral incisor pushed the mesh plate vertically, and the mucosa covering the cleft consequently swelled and thinned. A mucosal cut was made in the crest of the alveolar ridge over these teeth, and the part with the plate overlying the teeth was removed under local anesthesia. The canine and the lateral incisor then erupted approximately at the same time (Fig. 21).

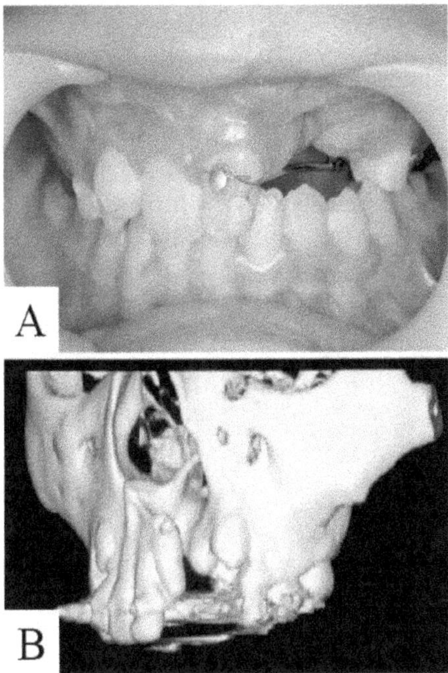

Fig. 20. The left unilateral cleft of the alveolus of our 9-year-old female patient. A: Intraoral view. B: Three-dimensional computed tomogram demonstrating the left maxillary canine and the alveolar bony defect (From Hibi et al. 2006. Reprinted with permission).

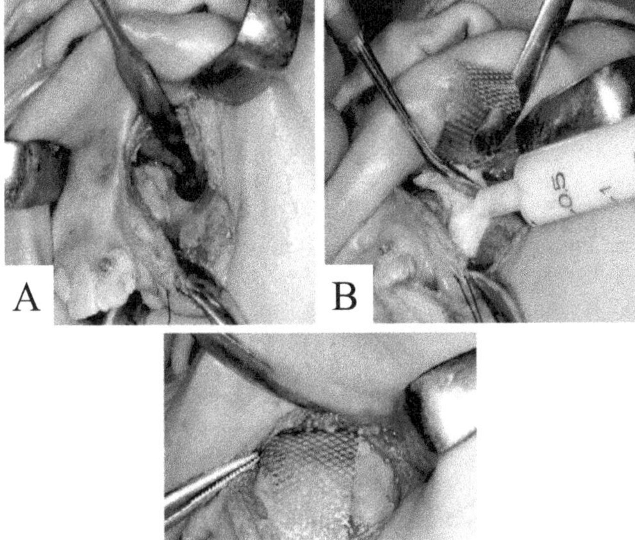

Fig. 21. Intraoperative views: A: Exposed alveolar cleft defect. B: Cleft cavity grafted with the tissue-engineered osteogenic material. C: Graft covered with the titanium mesh plate (From Hibi et al. 2006. Reprinted with permission).

This material regenerated the bone in the alveolar cleft defect without donor-site morbidity resulting from the autologous bone graft. Grafted bone remodels new bone due to apposition following resorption, and Van der Meij et al. reported that 1-year postoperative volumetric rates were approximately 70% for secondary bone grafts before canine eruption [66]. Using their measuring method [67] at 9 months postoperatively the present case showed 79.1% regenerated bone. The same authors indicated that the eruption of the canine generally occurred 2 years after bone graft if the patient was 9 years old. High resorbability of the bone in the grafted region may result in the early eruption of canine (Fig. 22).

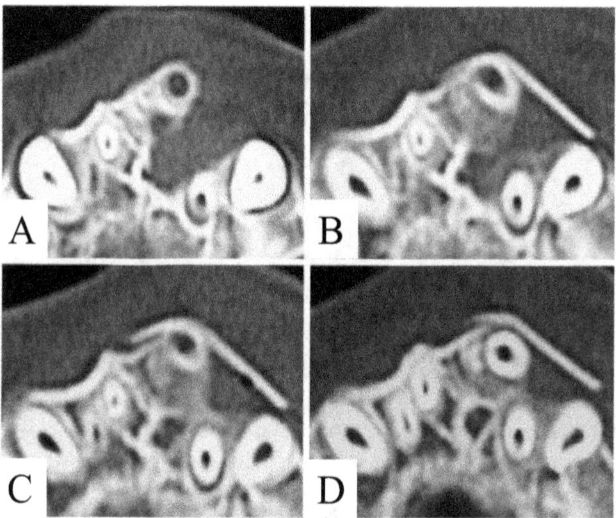

Fig. 22. Serial computed tomograms slicing the middle level of the alveolar cleft. A: Preoperation. B: Three months postoperation. Dome-shaped radiopaque images facing together and extending from the cleft bony walls inside the cavity. C: Six months postoperation. Fused image in the cleft cavity. D: Nine months postoperation. The lateral and supernumerary incisors are approximated in the bony bridge lateral to the reconstructed incisive canal (From Hibi et al. 2006. Reprinted with permission).

In the present case the canine coronally forced the mesh plate at 9 months postoperatively, which was earlier than expected. As the bone regenerated in the cleft defect, the ingrowing bone seemed to accompany the roots of not only the canine but also the lateral and supernumerary incisors, which consequently approximated and erupted. Bone regeneration with the injectable tissue-engineered may therefore have helped to induce teeth to reposition properly in the horizontal and vertical planes. Distraction of the transport bony segment has been attempted for closing alveolar defects. The defects are actually only reduced and not eliminated, and the teeth in the transport segment also were moved unintentionally according to the distraction. Some alteration in teeth positions may be beneficial, but others compromise crown morphology or require its recontouring. The bone transport in repair of the alveolar cleft therefore remains controversial (Fig. 23). Injectable tissue-engineered bone thus has a promising future. Its repeatability will also facilitate the sequential treatments of cleft palate patients.

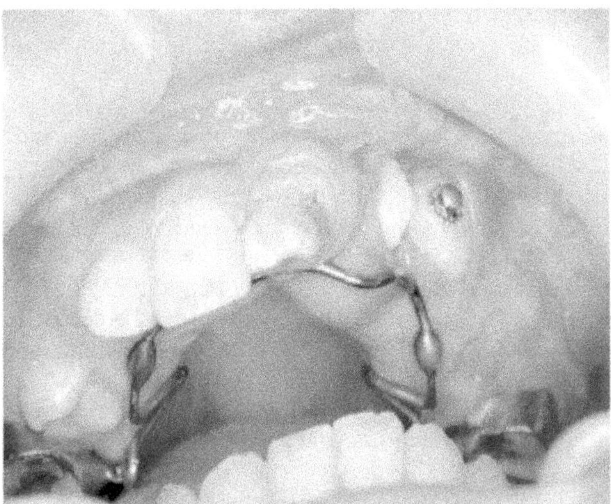

Fig. 23. The canine and the lateral incisor erupting in the reconstructed alveolar ridge (From Hibi et al. 2006. Reprinted with permission).

Taken together injectable tissue-engineered bone would provide a further option as a graft material for maxillary sinus floor augmentation, periodontal treatment, and distraction. The use of injectable tissue-engineered bone may well decrease healing time in days to come. Further, a tissue-engineered bone-induced increase in bone mass would potentially provide a great benefit to patients in cranio-maxillofacial and plastic surgery and to the bone reconstruction of other parts. Future research must address the long-term success rates of implants, the stability of tissue-engineered bone, and the application of the therapy to a less vascularized environment. Based on the present findings, future clinical trials are warranted.

References

1. Dahlin C, Sennerby L, Lekholm U, Linde A, Nyman S. Generation of new bone around titanium implants using a membrane technique: an experimental study in rabbits. Int J Oral Maxillofac Implants. 4: 19,1989
2. Hirsch JM, Ericsson I. Maxillary sinus augmentation using mandibular bone grafts and simultaneous installation of implants. A surgical technique. Clin Oral Implants Res. 2: 91,1991
3. Smiler DG, Johnson PW, Lozada JL, Misch C, Rosenlicht JL, Tatum OH Jr, Wagner JR. Sinus lift grafts and endosseous implants. Treatment of the atrophic posterior maxilla. Dent Clin North Am. 36: 151,1992
4. Moy PK, Lundgren S, Holmes RE. Maxillary sinus augmentation: histomorphometric analysis of graft materials for maxillary sinus floor augmentation. J Oral Maxillofac Surg. 51: 857,1993
5. Wheeler SL, Holmes RE, Calhoun CJ. Six-year clinical and histologic study of sinus-lift grafts. Int J Oral Maxillofac Implants. 11: 26,1996

6. Wood RM, Moore DL. Grafting of the maxillary sinus with intraorally harvested autogenous bone prior to implant placement. Int J Oral Maxillofac Implants. 3: 209,1988

7. Yamada Y, Boo JS, Ozawa R, Nagasaka T, Okazaki Y, Hata K, Ueda M. Bone regeneration following injection of mesenchymal stem cells and fibrin glue with a biodegradable scaffold. J Craniomaxillofac Surg. 31: 27,2003

8. Yamada Y, Ueda M, Naiki T, Takahashi M, Hata K, Nagasaka T. Autogenous injectable bone for regeneration with mesenchymal stem cells and platelet-rich plasma. Tissue-engineered bone regeneration. Tissue Eng. 10: 955,2004

9. Langer R, Vacanti JP. Tissue engineering. Science. 14: 920,1993

10. Asahara T, Murohara T, Sullivan A, Silver M, van der Zee R, Li T, Witzenbichier B, Schatteman G, Isner JM. Isolation of putative progenitor endothelial cells for angiogenesis. Science. 14: 964,1997

11. Kadiyala S, Young RG, Thiede MA, Bruder SP. Culture expanded canine mesenchymal stem cells possess osteochondrogenic potential *in vivo* and *in vitro*. Cell Transplant. 6: 125,1997

12. Boo JS, Yamada Y, Okazaki Y, Hibino Y, Okada K, Hata K, Yoshikawa T, Sugiura Y, Ueda M. Tissue-engineered bone using mesenchymal stem cells and a biodegradable scaffold. J Craniofac Surg. 13: 231,2002

13. Isogai N, Landis WJ, Mori R, Gotoh Y, Gerstenfeld LC, Upton J, Vacanti JP. Experimental use of fibrin glue to induce site-directed osteogenesis from cultured periosteal cells. Plast Reconstr Surg. 105: 953,2000

14. Bösch P, Braun F, Eschberger J, Kovac W, Spängler HP. The action of high-concentrated fibria on bone healing. Arch Orthop Unfallchir. 29: 259,1977

15. Schwarz N. The role of fibrin sealant in osteoinduction. Ann Chir Gynaecol Suppl. 207: 63,1993

16. Schlag G, Redl H, Schwarz N, Schiesser A, Lintner F, Dinges HP, Thurnher M. The influence of fibrin sealant on demineralized bone matrix-dependent osteoinduction. A quantitative and qualitative study in rats. Clin Orthop Relat Res. 238: 282,1989

17. Ilizarov GA. The tension-stress effect on the genesis and growth of tissues: Part II. The influence of the rate and frequency of distraction. Clin Orhtop Relat Res. 239: 263,1989

18. Swennen G, Dempf R, Schliephake H. Cranio-facial distraction osteogenesis: a review of the literature. Part II: experimental studies. Int J Oral Maxillofac Surg. 32: 123,2002

19. Takushima A, Kitano Y, Harii K. Osteogenic potential of cultured periosteal cells in a distraction gap in rabbits. J Surg Res. 78: 68,1998

20. Tsubota S, Tsuchiya H, Shinokawa Y, Tomita K, Minato H. Transplantation of osteoblast-like cells to distracted callus in rabbits. J Bone Joint Surg Br. 81B: 125,1999

21. Richards M, Huibregtse BA, Caplan AI, Goulet JA, Goldstein SA. Marrow-derived progenitor cell injections enhance new bone formation during distraction. J Orthop Res. 17: 900,1999

22. Takamine Y, Tsuchiya H, Kitakoji T, Kurita K, Ono Y, Ohshima Y, Kitoh H, Ishiguro N, Iwata H. Distraction osteogenesis enhanced by osteoblast-like cells and collagen gel. Clin Orthop. 399: 240,2002

23. Yamada Y, Ueda M, Hibi H, Nagasaka T. Translational research for injectable tissue-engineered bone regeneration using mesenchymal stem cells and platelet-rich plasma: From basic research to clinical application. Cell Transplant. 13: 343,2004

24. Squier CA, Ghoneim S, Kremenak CR. Ultrastructure of the periosteum from membrane bone. J Anat. 171: 233,1990

25. Tenenbaum HC, Heersche JN. Dexamethasone stimulates osteogenesis in chick periosteum *in vitro*. Endocrinology. 117: 2211,1985

26. Zohar R, Jaro S, Christopher A, McCulloch G. Characterization of stromal progenitor cells enriched by flow cytometry. Blood. 90: 3471,1997

27. Mizuno H, Hata K, Kojima, K, Bonassar LJ, Vacanti CA, Ueda M. A novel approach to regenerating periodontal tissue by grafting autologous cultured periosteum. Tissue Eng. 12: 1227,2006

28. Mase J, Mizuno, H, Okada K, Sakai K, Mizuno D, Usami K, Kagami H, Ueda M. Cryopreservation of cultured periosteum: effect of different cryoprotectants and pre-incubation protocols on cell viability and osteogenic potential. Cryobiology. 52: 182,2006

29. Ripamonti U, Crooks J, Petit JC, Rueger DC. Periodontal tissue regeneration by combined applications of recombinant human osteogenic protein-1 and bone morphogenetic protein-2. A pilot study in Chacma baboons (Papio ursinus). Eur J Oral Sci. 109: 241,2001

30. Jin QM, Anusaksathien O, Webb SA, Rutherford RB, Giannobile WV. Gene therapy of bone morphogenetic protein for periodontal tissue engineering. J Periodontol. 74: 202,2003

31. Taba M Jr, Jin Q, Sugai JV, Giannobile WV. Current concepts in periodontal bioengineering. Orthod Craniofac Res. 8: 292,2005

32. Yamada Y, Nakamura S, Ito K, Kohgo T, Hibi H, Nagasaka T, Ueda M. Injecctable tissue-engineered bone using autogenous bone marrow-derived stromal cells for maxillary sinus augmentation: clinical application report from a 2-6 year follow-up. Tissue Eng. A.14: 1699,2008

33. Yamada Y, Ueda M, Hibi H, Baba S. A novel approach to periodontal tissue regeneration with mesenchymal stem cells and platelet-rich plasma using tissue engineering technology: a clinical case report. Int J Periodontics Restorative Dent. 26: 363,2006

34. Hibi H, Yamada Y, Ueda M, Endo Y. Alveolar cleft osteoplasty using tissue-engineered osteogenic material. Int J Oral Maxillofac Surg. 35: 551,2006

35. Tatum H. Maxillary and sinus implant reconstructions. Dent Clin North Am. 30: 207,1986

36. Jensen J, Simonsen EK, Sindet-pedersen S. Reconstruction of the severely resorbed maxilla with bone grafting and osseointegrated implants: a preliminary report. J Oral Maxillofac Surg. 48: 27,1990

37. Velich N, Nemeth Z, Toth C, Szabo G. Long-term results with different bone substitute used for sinus floor elevation. J Craniofac Surg. 15: 38,2004

38. Del FM, Testori T, Francetti L, Weinstein R. Systematic review of survival rates for implants placed in the grafted maxillary sinus. Int J Periodontics Restorative Dent. 24: 565, 2004

39. Esposito M, Grusovin MG, Worthington HV, Coulthard P. Interventions for replacing missing teeth: bone augmentation techniques for dental implant treatment. Cochrane Database Syst Rev. 25: CD003607,2006

40. Jensen OT, Shulman LB, Block MS, Iacono VJ. Report of the sinus consensus conference of 1996. Int J Oral Maxillofac Implants. 13: 11,1998

41. Ferrigno N, Laureti M, Fanali S, Grippaudo G. A long-term follow-up study of nonsubmerged ITI implants in the treatment of totally edentulous jaws Part I: Ten-year life table analysis of a prospective multicenter study with 1286 implants. Clin Oral Implants Res. 13: 260,2002

42. Ohya M, Yamada Y, Ozawa R, Ito K, Takahashi M, Ueda M. Sinus floor elevation applied for tissue-engineered bone – comparative study between mesenchymal stem cells (MSCs) and platelet-rich plasma (PRP) and autogenous bone with PRP complexes in rabbits. Clin Oral Implants Res. 16: 622,2005

43. Watanabe K, Niimi A, Ueda M. Autogenous bone grafts in the rabbit maxillary sinus. Oral Radiol Endod. 88: 26,1999

44. Ito K, Yamada Y, Nagasaka T, Baba H, Ueda M. Osteogenic potential of injectable tissue-engineered bone: a comparison among autogenous bone, bone substitute (Bio-oss®), platelet-rich plasma (PRP), and tissue-engineered bone with respect to their mechanical properties and histological findings. J Biomed Mater Res. A. 73: 63,2005

45. Raghoebar GM, Schortinghuis J, Liem RSB, Ruben JL, Wal JE, Vissink A. Does platelet-rich plasma promote remodeling of autogenous bone grafts used for augmentation of the maxillary sinus floor? Clin Oral. Implants Res. 16: 349,2005

46. Wada K, Niimi A, Sawaki T, Ueda M. Maxillary sinus floor augmentation in rabbits: a comparative histologic-histomorphometric study between rhBMP-2 and autogenous bone. Int J Periodontics Restorative Dent. 21: 253,2001

47. Marx RE, Carlson ER, Eichstaedt RM, Schimmele SR, Stauss JE, Georgeff KR. Platelet-rich plasma, growth factor enhancement for bone grafts. Oral Surg. Oral Med Oral Pathol Oral Radiol Endod. 85: 638,1998

48. Pihlstrom BL, McHugh RB, Oliphant TH, Ortiz-Campos C. Comparison of surgical and nonsurgical treatment of periodontal disease. J Clin Periodontal. 10: 524,1983

49. Knowles JW, Burgett FG, Nissle RR, Shick RA, Morrison EC, Ramfjord SP. Results of periodontal treatment related to pocket depth and attachment level. Eight years. J Periodontol. 50: 225,1979

50. Nyman S, Rosling B, Lindhe J. Effect of professional tooth cleaning on healing after periodontal surgery. J Clin Periodontol. 2: 80,1975

51. Karring T, Nyman S, Gottlow J, Laurel L. Development of the biological concept of guided tissue regeneration: Animal and human studies. Periodontology. 1: 26,1993

52. Gottlow J, Nyman S, Lindhe J, Karring T, Wennstrom J. New attachment formation in the human periodontium by guided tissue regeneration. Case Reports. J Clin Periodontol. 13: 604,1986

53. Pontoriero P, Lindhe J, Nyman S, Karring T, Rosenberg S. Guided tissue regeneration in degree II furcation-involved mandibular molars. Clinical study. J Clin Periodontol. 15: 247,1988

54. Cortellini P, Pini PG, Tonetti M. Periodontal regeneration of human infrabony defects. II. Re-entry procedures and bone measures. J. Periodontol. 64: 261,1993

55. Schroeder H. Biological problems of regenerative cementogenesis: Synthesis and attachment of collagenous matrices on growing and established root surfaces. Int Rev Cytol. 142: 1,1992

56. Araujo M, Berglundh T, Lindhe J. The periodontal tissues in healed degree III furcation defects. A experimental study in dogs. J Clin Periodontol. 23: 532,1996

57. Heijl L, Heden G, Svardstrom G, Ostgren A. Enamel matrix derivative (EMDOGAIN) in the treatment of infrabony periodontal defects. J Clin Periodontol. 24: 705,1997

58. Zecchelli G, Bernardi F, Montebugnoli L, De Sanctis M. Enamel matrix proteins and guided tissue regeneration with titanium-reinforced expanded polytetrafluoroethylene membranes in the treatment of infrabony defects: A coparative controlled clinical trial. J Periodontol. 73: 3,2002

59. De Sanctis M, Zucchelli G, Clauser C. Bacterial colonization of barrier material and periodontal regeneration. J Clin Periodontol. 23: 1039,1996

60. Nowzari H, Macdonald ES, Flynn J, London RM, Morrison JL, Slots J. The dynamics of microbial colonization of barrier membranes for guided tissue regeneration. J Periodontol. 67: 694,1996

61. Cortellini P, Tonetti MS. Focus on infrabony defects: Guided tissue regeneration. Periodontology. 22: 104,2000

62. Horswell BB, Henderson JM. Secondary osteoplasty of the alveolar cleft defect. J. Oral Maxillofac Surg. 61: 1082,2003

63. Zeitler D. Alveolar cleft grafts. In: Fonseca RJ, ed: Oral and Maxillofacial Surgery, Vol. 6 Cleft/ craniofacial/cosmetic surgery. Philadelphia: W.B. Saunders 75,2000

64. Fonseca RJ, Turvey TA, Wolford LM. Orthognathic surgery in the cleft patients In: Fonseca RJ, ed: Oral and Maxillofacial Surgery, Vol. 6 Cleft/ craniofacial/cosmetic surgery. Philadelphia: W.B. Saunders 87,2000

65. Hibi H, Yamada Y, Kagami H, Ueda M. Distraction osteogenesis assisted by tissue engineering in an irradiated mandible: a case report. Int J Oral Maxillofac Implants. 21: 141,2006

66. Van der Meij AJW, Baart JA, Prahl-Andersen B, Valk J, Kostense PJ, Tuinzing DB. Bone volume after secondary bone grafting in unilateral and bilateral clefts determined by computed tomography scans. Oral Surg Oral Med Oral Pathol Oral Radiol Endod. 92: 136,2001

67. Yen SL-K, Yamashita DD, Gross J, Meara JG, Yamazaki K, Kim TH, Reinisch J. Combining orthodontic tooth movement with distraction osteogenesis to close cleft spaces and improve maxillary arch form in cleft lip and palate patients. Am J Orthod Dentofacial Orthop. 127: 224,2005

(Ito K, Yamada Y, Naiki T, Usami K, Mizuno H, Okada K, Narita Y, Aoki M, Kondo T, Mizuno D, Mase J, Nishiguchi H, Kagami H, Kinoshita K, Hibi H, Nagasaka T, Ueda M)

Chapter 4

Cartilage

Cartilage regeneration using chondrocyte

Cartilage tissue engineering comprises three factors: cell source, growth factors, and scaffolds. Chondrocytes from other cartilage such as rib cartilage are most commonly used for the formation of cartilage tissue [1]. However, their cell number is limited and it is difficult to construct a tissue of large size. Differentiation of embryonic stem cells toward chondrocytes has been accomplished, but its clinical application is impractical at present from ethical points of view [2, 3]. In contrast, MSCs are promising because they can easily be prepared from patients without invasive surgery. These cells grow rapidly, retaining their capacity to differentiate into chondrocytes under certain conditions [4, 5]. Several growth factors such as TGF-β, BMPs, FGFs, and IGFs are involved in chondrocyte differentiation, proliferation, and maintenance [6]. These molecules are used for cartilage tissue engineering. The application of scaffolds has two advantages in this type of engineering. It enables three-dimensional culture, which is a necessary microenvironment for maturing the chondrocyte phenotype. It serves as an artificial matrix that gradually becomes replaced with native cartilage matrix. Although several methods have been attempted, with consideration of these factors, no cartilage tissue has been engineered that fulfills clinical requirements. There has been accumulating evidence that stimulation of chondrocytes facilitates cartilage matrix formation [7].

For instance, hydrostatic pressure on bovine chondrocytes is known to enhance their matrix synthesis and accumulation [8, 9]. Direct compression on bovine chondrocytes embedded in agarose gel increases glycosaminoglycan and collagen composition [10]. Thus, stress may serve as another important factor in cartilage tissue engineering. In clinical studies and animal models, low-intensity ultrasound (US) promotes fracture repair and increases mechanical strength. US also promotes cartilage healing by increasing glycosaminoglycan synthesis of chondrocytes. As MSCs have the ability to differentiate into chondrocytes, US may promote their differentiation. Here, we evaluated the effects of US on the differentiation of MSCs toward chondrocytes and cartilage matrix formation. When human MSCs cultured in pellets were treated with TGF-β (10 ng/ml), they differentiated into chondrocytes as assessed by alcian blue staining and immunostaining for aggrecan, but nontreated cell pellets did not. Furthermore, when low-intensity US was applied for 20 minutes every day to the TGF-β-treated cell pellets, chondrocyte differentiation was enhanced (Fig. 24).

Biochemically, aggrecan deposition was increased by 2.9- and 8.7-fold by treatment with TGF-β alone, and with both TGF-β and US, respectively. In contrast, cell proliferation and total protein amount appeared unaffected by these treatments. These results indicate that low-intensity US enhances TGF-β-mediated chondrocyte differentiation of MSCs in pellet culture and that application of US may facilitate larger preparations of chondrocytes and the formation of mature cartilage tissue.

MSCs had to be cultured in pellets for differentiation into chondrocytes. Even in the presence of TGF-β, they did not differentiate when cultured on plates. Induction of chondrocyte differentiation in MSCs in pellets may imply a requirement for cell–cell interaction different from that in plate culture, as confluent cells show little differentiation, and the pellet culture may provide a microenvironment similar to mesenchymal condensation, which normally takes place on initiation of chondrogenesis [11]. It has been demonstrated that chondrocytes maintain differentiation in pellets 12 or in three-dimensional culture coupled with scaffolds such as alginate beads, 34 collagens, and polyglycolic acid [12, 13]. The cells in the pellet may have an appropriate microenvironment for differentiation. Studies on the expression patterns of MSCs cultured in pellets during chondrocyte differentiation demonstrate that these cells exhibit sequential expression of molecules involved in chondrocyte differentiation. The pellet culture system of MSCs will enable us to study US effects at different stages of chondrocyte differentiation.

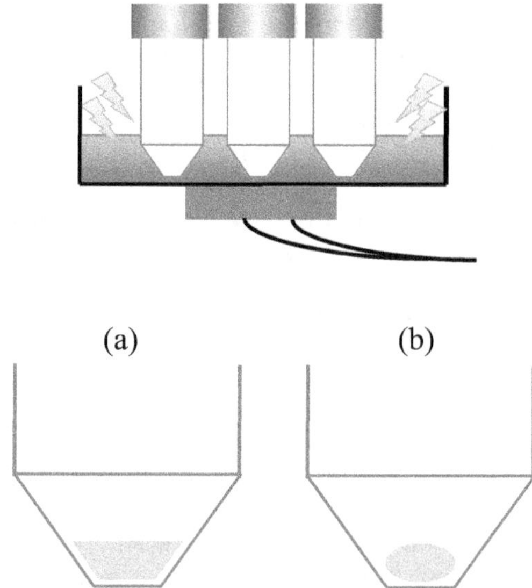

Fig. 24. Schematic drawing of the application of ultrasound and the pellet culture. (a) Cell pellet of hMSCs immediately after centrifugation appears quadrilateral. (b) The cell pellet after 24 hours of incubation appears oval (From Ebisawa et al. 2004).

References

1. Brittberg M, Lindahl A, Nilsson A, Ohlsson C, Isakson O, Peterson L. Treatment of deep cartilage defects in the knee with autologous chondrocyte transplantation. N Engl J Med. 331: 889,1994
2. Kramer J, Hegert C, Guan K, Wobus AM, Müller PK, Rohwedel J. Embryonic stem cell-derived chondrogenic differentiation *in vitro*: activation by BMP-2 and BMP-4. Mech Dev. 92: 193,2000
3. Paul G, Li JY, Brundin P. Stem cells: Hype or hope? Drug Discov Today. 7: 295,2002
4. Johnstone B, Hering TM, Caplan AI, Goldberg VM, Yoo JU. *In vitro* chondrogenesis of bone marrow-derived mesenchymal progenitor cells. Exp Cell Res. 10: 238,1998

5. Yoo JU, Barthel TS, Nishimura K, Solchaga L, Caplan AI, Goldberg VM, Johnstone B. The chondrogenic potential of human bone-marrow-derived mesenchymal progenitor cells. J Bone Joint Surg Am. 80: 1745,1998

6. Raddi AH. Symbiosis of biotechnology and biomaterials: applications in tissue engineering of bone and cartilage. J Cell Biochem. 56: 192,1994

7. Darling EM, Athanasiou KA. Articular cartilage bioreactors and bioprocesses. Tissue Eng. 9: 9,2003

8. Mizuno S, Tateishi T, Ushida T, Glowacki J. Hydrostatic fluid pressure enhances matrix synthesis and accumulation by bovine chondrocytes in three-dimensional culture. J Cell Physiol. 193: 319,2002

9. Ikenoue T, Trindade MC, Lee MS, Lin EY, Schurman DJ, Goodman SB, Smith RL. Mechanoregulation of human articular chondrocyte aggrecan and type II collagen expression by intermittent hydrostatic pressure *in vitro*. J Ortho Res. 21: 110,2003

10. Buschmann MD, Gluzband YA, Grodzinsky AJ, Hunziker EB. Mechanical compression modulates matrix biosynthesis in chondrocyte/agarose culture. J Cell Sci. 108: 1497,1995

11. Thorogood PV, Hinchliffe JR. An analysis of the condensation process during chondrogenesis in the embryonic chick hind limb. J Embryol Exp Morphol. 33: 581,1975

12. Guo JF, Jourdian GW, MacCallum DK. Culture and growth characteristics of chondrocytes encapsulated in alginate beads. Connect Tissue Res. 19: 277,1989

13. Wakitani S, Kimura T, Hirooka A, Ochi T, Yoneda M, Yasui N, Owaki H, Ono K. Repair of rabbit articular surfaces with allograft chondrocytes embedded in collagen gel. J Bone Joint Surg Br. 71: 74,1989

(Ebisawa K, Hata K, Okada K, Kimata K, Torii S, Watanabe H, Ueda M)

Chapter 5

Tooth

Tissue-engineered odontogenesis

One important goal of dental research is the efficient regeneration of lost teeth [1, 2]. Tooth formation, or odontogenesis, is a complex process that has been characterized as a series of reciprocal epithelial-mesenchymal interactions, culminating in the differentiation of the interacting tissues [3-6]. Tissue engineering of tooth structures on biodegradable polymer scaffolds has been recently achieved [7]. The method involves dissociating porcine third molar tooth buds into single-cell suspensions and seeding them onto biodegradable polymers, but this regeneration process is not yet fully understood. To characterize the process in greater detail, we followed the regeneration of tissue-engineered teeth by histology and immunohistochemistry with specific markers of epithelial and mesenchymal differentiation as well as ameloblasts, odontoblasts, and cementoblasts. In the present study, we show that the development of these engineered teeth closely parallels that of natural odontogenesis.

Preparation of biodegradable polymer scaffolds

Three-dimensional scaffolds were prepared as described previously [8], with the following modifications: polyglycolic acid (PGA) fiber mesh (fiber diameter = 13 µm; density = 60 mg/ml) was packed into 96-well plates and sterilized in 75% ethanol. The scaffold dimensions were approximately 1 cm³. Before seeding the cells, the scaffolds were collagen-coated overnight at 4°C (1 mg/ml type I collagen in 10 mM HCl), followed by three times washings in PBS and three times in DMEM.

Isolation and dissociation of porcine third molar tissue

Tooth bud cells were harvested and prepared as described previously [7]. The tooth buds were removed early in development at the early stage of crown formation from the mandibles of 6-month-old pigs, which were purchased from a local slaughterhouse and transported to the laboratory on ice. Briefly, third molar tooth buds were removed from the fresh mandibles with their inner and outer epithelial layers as well as the dental papilla and dental follicle intact (Fig. 25). After the calcified tissue was removed, all tooth bud tissue, including the dental follicles, was minced into <1 mm³ pieces in Hanks Balanced Salt Solution (HBSS) and dissociated with 2 mg/ml collagenase and dispase for 50 minutes at 37°C, followed by gentle trituration. Digested tooth bud tissues were then strained through a nylon filter (70 µm pores), and the isolated single cells (1.0 × 10⁷ cells) were seeded onto a PGA fiber mesh scaffold which had been precoated with type I collagen (1 mg/ml type I collagen in 10 mM HCl) for 3 hours at 4°C before placing them into the omentum in the host rats (Fig. 26).

The animals were anesthetized with an intraperitoneal injection of sodium pentobarbital (15 mg/kg), and the scaffolds seeded with cells were implanted into the omenta of the athymic rats (n = 30) [9]. Samples developed for 2, 4, 6, 8, 10, 15, 20, and 25 weeks were dissected and immediately fixed in freshly prepared 4%   paraformaldehyde in PBS at 4°C for 6-8 hours. After fixation, the tissues were demineralized for 4 h in 0.2N HCl and, after extensive washing in PBS, were dehydrated in an ethanol gradient, cleared in xylene, and embedded in paraffin. Tissue sections 6 μm thick were mounted on glass slides and stained with hematoxylin and eosin (H-E)   .

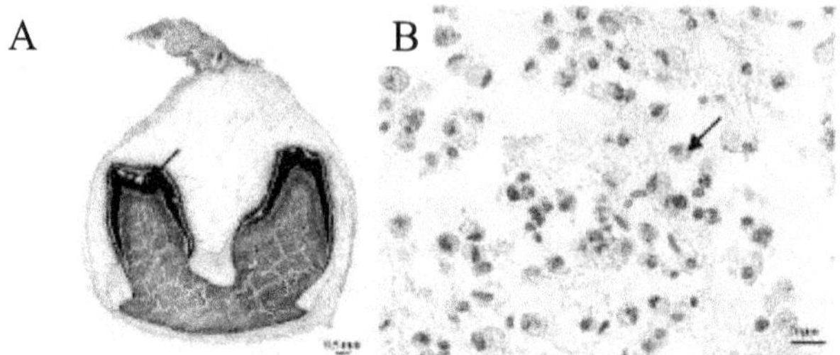

Fig. 25. A: A third molar tooth bud used in experiments stained with toluidine blue. The cusps of the molar are already calcified (black arrow). B: Single cells dissociated by straining through a nylon filter (70-μm pores) before seeding onto the PGA fiber mesh scaffold (black arrow) (From Honda et al. 2005. Reprinted with permission).

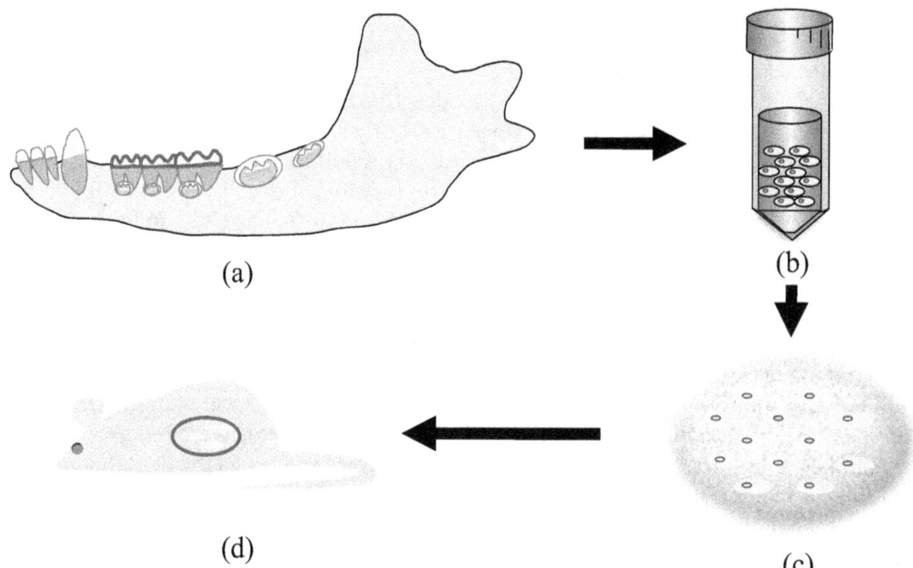

Fig. 26. Schematic diagram of the strategy used to produce the tissue-engineered tooth. Tooth-germ was derived from isolated cells seeded on a PGA fiber mesh scaffold. (a) Tissue was harvested from third molar tooth buds. (b) Isolate the tooth bud. (c) Isolated cells were seeded on PGA mesh. (d) Implantation in rat omentum (From Honda et al. 2005).

In vivo implantation and histology
Aggregates of epithelial cells were first observed 4-6 weeks after implantation (Fig. 27). These aggregates assumed three different shapes: a natural tooth germ-like shape, a circular shape, or a bilayer-bundle. Based on the structure of the stellate reticulum in the dental epithelium, the circular and bilayer-bundle aggregates could be clearly classified into two types: one with extensively developed stellate reticulum, and the other with negligible stellate reticulum. The epithelial cells in the circular aggregates differentiated into ameloblasts. The continuous bilayer bundles eventually formed the epithelial sheath, and dentin tissue was evident at the apex of these bundles. Finally, enamel-covered dentin and cementum-covered dentin formed, a process most likely mediated by epithelial-mesenchymal interaction. These results suggest that the development of these engineered teeth closely parallels that of natural odontogenesis derived from the immature epithelial and mesenchymal cells.

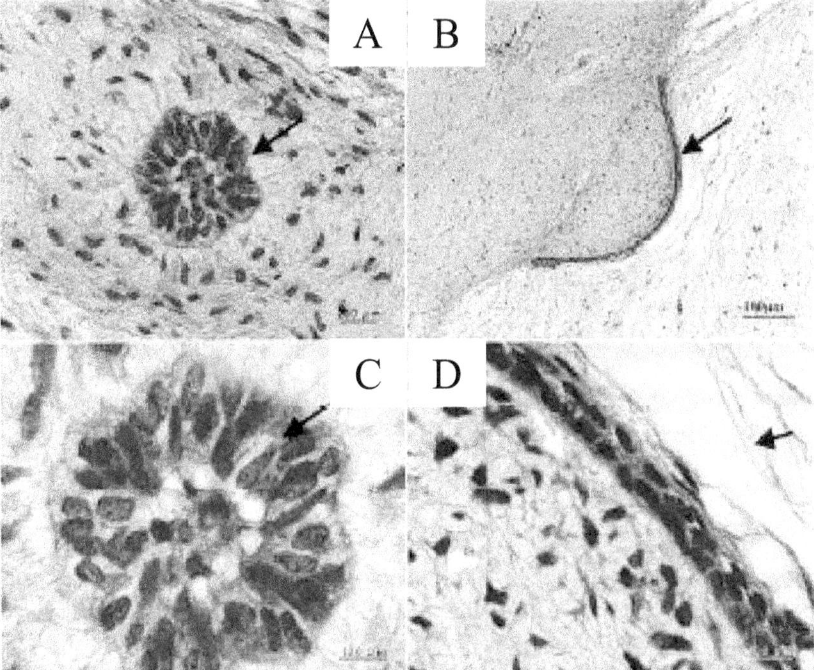

Fig. 27. A, B: At 6-8 weeks after implantation. A: The cells have formed either circular aggregates (black arrow). B: Sheets of a bilayer of cells (black arrow) which separate two very different densities of surrounding cells. C: At higher magnification, the cells in circular aggregates are low columnar in shape with nuclei in the center of the cytoplasm (black arrow). D: While the cells in the sheets clearly form a two-cell thick layer, which separates a low density of cells (black arrow) from a denser region of cells (white arrowhead), which is densest just adjacent to the bilayer (white arrow) (From Honda et al. 2005. Reprinted with permission).

Effects of shear stress
During tooth organogenesis, reciprocal interactions between epithelial-mesenchymal cells result in the cytodifferentiation of epithelial cells into ameloblasts and ectomesenchymal cells into odontoblasts [5, 11-13]. These differentiated cells produce the enamel and the dentin matrix. Amelogenesis is the process of enamel formation which occurs in three distinct stages: the presecretory stage, consisting of ameloblast cytodifferentiation; the secretory stage, with the bulk of enamel matrix formation; and the maturation stage, associated with matrix mineralization [14-16]. Dentinogenesis is the process of dentin formation, by which the dental papilla cells located at the epithelial-mesenchymal interface gradually differentiate into odontoblasts concomitant with ameloblast differentiation [17-20]. Recently, several *in vitro* and *in vivo* studies have demonstrated that dental pulp cells are also capable of differentiating into odontoblasts and producing a mineralizing matrix [21]. There is accumulating evidence that mechanical stress has a variety of effects on cell growth and differentiation [22, 23]. For example, such stress is known to facilitate the differentiation of osteoblasts [24] and chondrocytes [25]. In particular, shear stress induced by fluid flow facilitates the secretion of bone matrix protein [24, 26]. It is demonstrated that subjecting osteoblasts to fluid shear stress increases expression of genes including c-fos and cyclooxygenase-2 (COX-2) [27]. We hypothesized that appropriate shear stress is essential to facilitate the differentiation of odontogenic cells, which would facilitate the regeneration process. The purpose of this study was to investigate the effect of applying shear stress on the differentiation of odontogenic cells and histogenesis of odontogenic tissues. This study shows for the first time that shear stress facilitates tissue-engineered odontogenesis. Furthermore, using RT-PCR and Western blot technique, we provide evidence for the expression of teeth-related marker as to whether tooth cells facilitate cell differentiation by exposure to shear stress.

Shear stress exposure
In preliminary studies, we examined the effects of three different types of mechanical stress on the differentiation of the cells. Uniaxial stretch, ultrasonic wave, and shear stress generated by bi-directional fluid flow were given, and the levels of alkaline phosphatase (ALP) activity were evaluated as a marker for pulp cellular differentiation [28, 29]. Among the mechanical stresses examined, only shear stress succeeded in increasing ALP activity. From this result, the shear stress appeared to influence most clearly the cells on the ALP activity. Shear stress was generated by bi-directional fluid flow inside a tube, the velocity of which depended on the frequency of the agitating motion of the Shaker. After cells were seeded onto the scaffolds, cell-polymer constructs (CPCs) were first cultured for 2 hours under static conditions. Subsequently, CPCs were placed into 15-ml centrifuge tubes with DMEM containing 10% of fetal calf serum (FCS). To determine the more effective stress, we compared three distinct shear stresses by exposing seeding cells and evaluated the ALP activity. The used frequencies were between 10-20, 40-50, and 70-80 rpm at 37°C for 12 hours. Swing frequencies of constructs were 40, 90, and 120 rpm, respectively. Their mechanical load exerted on the CPCs was estimated at 0.7×10^{-4} N, 1.2×10^{-4} N, and 1.6×10^{-4} N, respectively. *In vitro* studies, the expression of both epithelial and mesenchymal odontogenic-related mRNAs was significantly enhanced by shear stress for 2 hours. Twelve hours after exposure to shear stress, the expression of amelogenin, bone sialoprotein and vimentin protein was significantly enhanced compared with that of control. After 7 days,

alkaline phosphatase activity exhibited a significant increase without any significant effect on cell proliferation *in vitro*. *In vivo*, enamel and dentin tissues formed after 15 weeks of *in vivo* implantation in constructs exposed to *in vitro* shear stress for 12 hours. Such was not the case in controls. We concluded that shear stress facilitates odontogenic cell differentiation *in vitro* as well as the process of tooth tissue engineering *in vivo*.

Collagen sponge as a 3-D scaffold

Tooth structure can be regenerated by seeding dissociated tooth cells onto PGA fiber mesh, although the success rate of tooth production is low. The present study was designed to compare the performance of collagen sponge with PGA fiber mesh as a 3-D scaffold for tooth-tissue engineering (Fig. 28). Porcine third molar teeth at the early stage of crown formation were enzymatically dissociated into single cells, and the heterogeneous cells were seeded onto collagen sponge or the PGA fiber mesh scaffolds. Scaffolds were then cultured to evaluate cell adhesion and ALP activity *in vitro*. An *in vivo* analysis was performed by implanting the constructs into the omentum of immunocompromised rats and evaluating tooth production up to 25 weeks (Fig. 29). After 24 hours, there were a significantly higher number of cells attached to the collagen sponge scaffold than the PGA fiber mesh scaffold. Similarly, the ALP activity was significantly higher for the collagen sponge scaffold was than the PGA fiber mesh scaffold after 7 days of culture. The area of calcified tissue formed in the collagen sponge scaffold was also larger than in the PGA fiber mesh scaffold (Fig. 30). The results from *in vivo* experiments show conclusively that a collagen sponge scaffold allows tooth production with a higher degree of success than PGA fiber mesh. Taken together, the results from this study show that collagen sponge scaffold is superior to the PGA fiber mesh scaffold for tooth-tissue engineering.

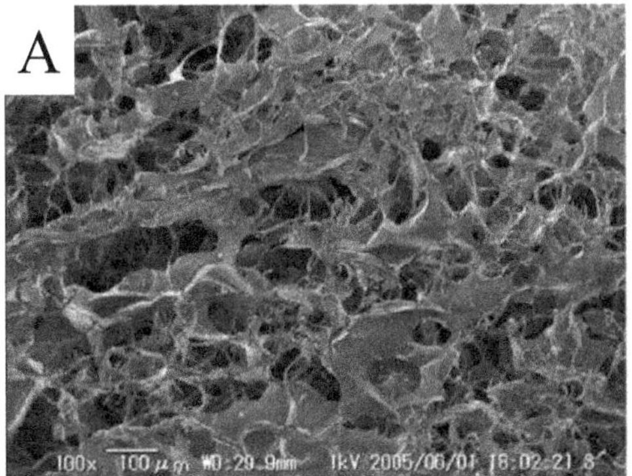
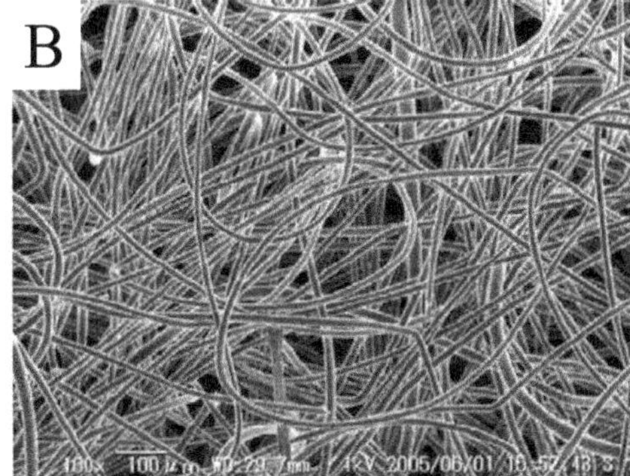

Fig. 28. SEM images of both scaffold types. A: Collagen sponge. B: PGA fiber mesh (From Sumita et al. 2006. Reprinted with permission).

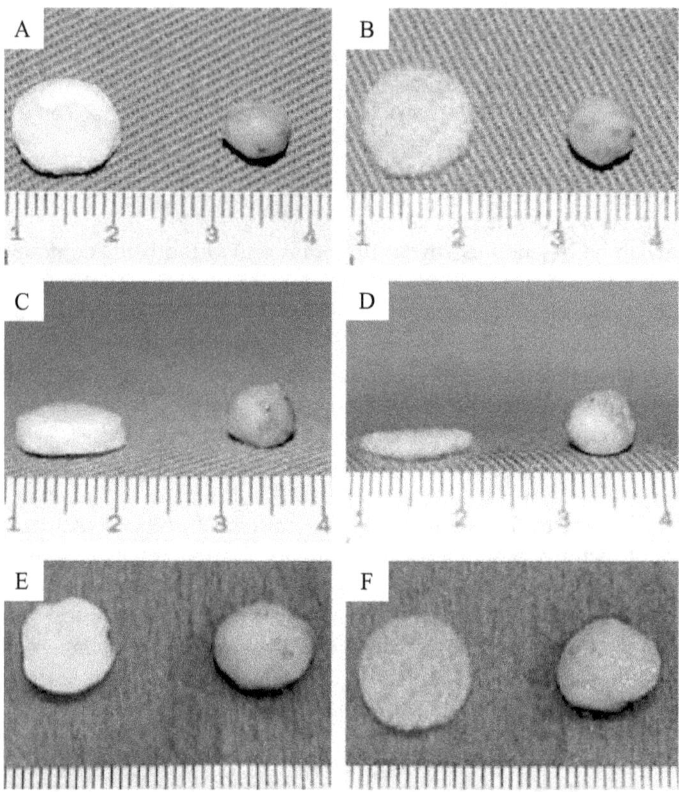

Fig. 29. Gross appearance of the original scaffold and implants after 8 and 20 weeks implantation. (A, C and E) show the collagen sponge scaffold before implantation and the implants obtained from the collagen sponge scaffold at 8 (A and C) and 20 weeks (E) after implantation. (B, D and F) show the PGA fiber mesh scaffold before implantation and the implants at 8 (B and D) and 20 weeks (F) after implantation. A: The original collagen sponge scaffold (left) and the implant at 8 weeks after implantation (right). The diameter of the original collagen sponge scaffold was approximately 11 mm, while the diameter of the implant was approximately 7 mm. B: The original PGA fiber mesh scaffold (left) and the implant at 8 weeks after implantation (right). The diameter of the original PGA fiber mesh scaffold was approximately 11 mm, while the diameter of the implant was approximately 7 mm. C: The perpendicular thickness of the original collagen sponge scaffold was approximately 2 mm (left), while the perpendicular thickness of the implant at 8 weeks was approximately 7 mm (right). D: The perpendicular thickness of the original PGA fiber mesh scaffold was approximately 1–2 mm (left), while the perpendicular thickness of the implants at 8 weeks was approximately 7 mm (right). E: The original collagen sponge scaffold (left) and the implant at 20 weeks after implantation. The diameters of the implant were approximately 11 mm by 11 mm by 10 mm (right). F: The original PGA fiber mesh scaffold (left) and the implant at 20 weeks after implantation. The diameters of the implants were approximately 11 mm by 11 mm by 10 mm (right) (From Sumita et al. 2006. Reprinted with permission).

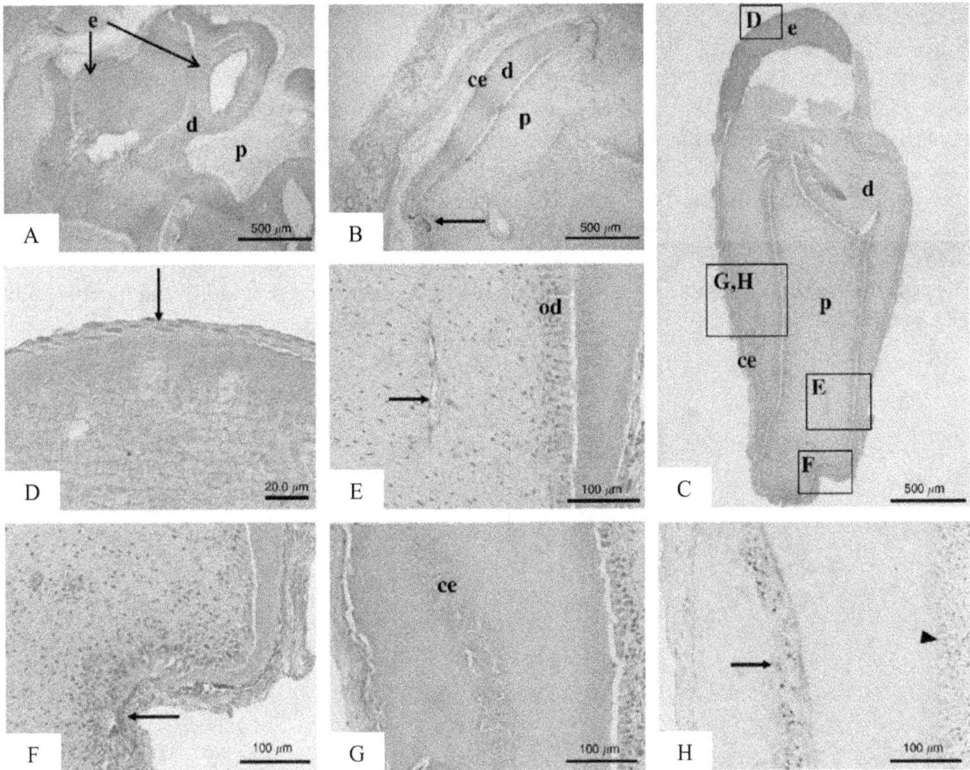

Fig. 30. Histological and immunohistochemical analyses of TE-teeth at 25 weeks after implantation. (A–G) show H-E staining and panel (H) shows immunohistochemical staining for BSP. A: The circular-shaped TE-teeth obtained from the collagen sponge scaffold revealed the enamel (e) and dentin (d). B: The stick-shaped TE-teeth obtained from the collagen sponge scaffold revealed the dentin (d), cementum-like tissue (ce), pulp (p) and HERS (black arrow). C: A typical tooth with normal morphology was observed in the implant obtained from the collagen sponge scaffold. TE-teeth revealed (e), (d), (p) and (ce). (D–H) are higher magnifications of (C) as indicated by the squares. D: The reduced enamel epithelium cells (black arrow) were recognized on the surface of the enamel. E: The odontoblasts (od) and blood vessels (black arrow) were identified in the pulp. F: Bilayers of epithelial cells were similar to Hertwig's epithelial root sheath (black arrow). G: Cellular cementum-like tissue (ce) was observed on the surface of the root dentin. H: BSP expression was located in cellular cementum-like tissue (black arrow) and odontoblasts (black arrowhead) (From Sumita et al. 2006. Reprinted with permission).

Sequential seeding

Progress is being made toward regenerating teeth by seeding dissociated postnatal odontogenic cells onto scaffolds and implanting them *in vivo*, but tooth morphology remains difficult to control. In this study, we aimed to facilitate tooth regeneration using a novel technique to sequentially seed epithelial cells and mesenchymal cells so that they developed appropriate interactions in the scaffold. Dental epithelium and mesenchyme from porcine third molar teeth were enzymatically separated and dissociated into single cells. Mesenchymal cells were seeded onto the surface of the scaffold and epithelial cells were then plated on top so that the two cell types were in direct contact. The cell-scaffold constructs were evaluated *in vitro* and also implanted into immunocompromised rats for *in vivo* analysis (Fig. 31). Control groups included constructs where direct contact between the two cell types was prevented. In scaffolds, seed using the novel technique, ALP activity, was significantly greater than controls, the tooth morphology *in vivo* was developed similar to that of a natural tooth, and only one tooth structure formed in each scaffold. These results suggest that the novel cell-seeding technique could be useful for regulating the morphology of regenerated teeth.

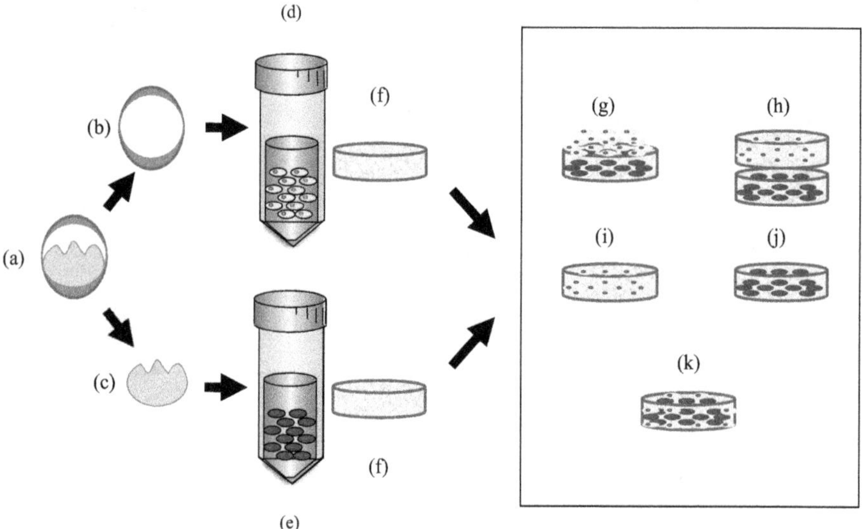

Fig. 31. Experimental design for the study.(a) Tooth germ. (b) Epithelium. (c) Mesenchyme. (d) Epithelial cell suspension. (e) Mesenchymal cell suspension. (f) Collagen sponge. (g) Group I, mesenchymal cells were plated onto the scaffold, and epithelial cells were then plated on top of the mesenchymal cells; (h) Group II, mesenchymal cells were plated onto a scaffold in the lower compartment, and epithelial cells onto a scaffold in the upper compartment of a dish divided by a microporous membrane; (i) Group III, epithelial cells alone were seeded into the scaffold; and (j) Group IV, mesenchymal cells alone were seeded into the scaffold. For the *in vivo* study, the groups were similar but with an additional (k) Group V, in which both epithelial cells and mesenchymal cells were plated randomly into each scaffold (From Honda et al. 2007. Reprinted with permission).

References

1. Earthman JC, Sheets CG, Paquette JM, Kaminishi RM, Nordland WP, Keim RG, et al. Tissue engineering in dentistry. Clin Plast Surg. 30: 621,2003
2. Chai Y, Slavkin HC. Prospects for tooth regeneration in the 21st century: a perspective. Microsc Res Tech. 60: 469,2003

3. Thesleff I. Interactions between the extracellular matrix and the cell surface determine tooth morphogenesis and the cellular differentiation of the dental mesenchyme. Ontogenez. 20: 341,1989

4. Thesleff I, Vaahtokari A, Kettunen P, Aberg T. Epithelial-mesenchymal signaling during tooth development. Connect Tissue Res. 32: 9,1995

5. Jernvall J, Thesleff I. Reiterative signaling and patterning during mammalian tooth morphogenesis. Mech Dev. 92: 19,2000

6. McCollum M, Sharpe PT. Evolution and development of teeth. J Anat. 199: 153,2001

7. Young CS, Terada S, Vacanti JP, Honda M, Bartlett JD, Yelick PC. Tissue engineering of complex tooth structures on biodegradable polymer scaffolds. J Dent Res. 81: 695,2002

8. Mikos AG, Bao Y, Cima LG, Ingber DE, Vacanti JP, Langer R. Preparation of poly (glycolic acid) bonded fiber structures for cell attachment and transplantation. J Biomed Mater Res. 27: 183,1993

9. Choi RS, Vacanti JP. Preliminary studies of tissue-engineered intestine using isolated epithelial organoid units on tubular synthetic biodegradable scaffolds. Transplant Proc. 29: 848,1997

10. Bancroft JD, Stevens A. Theory and practice of histological techniques. 3rd ed. Edinburgh ; New York: Churchill Livingstone;1990

11. Thesleff I. Tooth morphogenesis. Adv Dent Res. 9: 12,1995

12. Maas R, Bei M. The genetic control of early tooth development. Crit Rev Oral Biol Med. 8: 4,1997

13. Thesleff I, Sharpe P. Signalling networks regulating dental development. Mech Dev. 67: 111,1997

14. Zeichner-David M, Diekwisch T, Fincham A, Lau E, MacDougall M, Moradian-Oldak J, Control of ameloblast differentiation. Int J Dev Biol. 39: 69,1995

15. Smith CE, Nanci A. Overview of morphological changes in enamel organ cells associated with major events in amelogenesis. Int J Dev Biol. 39: 153,1995

16. Robinson C, Brookes SJ, Shore RC, Kirkham J. The developing enamel matrix: nature and function. Eur J Oral Sci. 106 Suppl 1: 282,1998

17. Linde A, Goldberg M. Dentinogenesis. Crit Rev Oral Biol Med. 4: 679,1993

18. Butler WT. Dentin matrix proteins and dentinogenesis. Connect Tissue Res. 33: 59,1995

19. Sasaki T, Garant PR. Structure and organization of odontoblasts. Anat Rec. 245: 235,1996

20. Smith AJ, Cassidy N, Perry H, Begue-Kirn C, Ruch JV, Lesot H. Reactionary dentinogenesis. Int J Dev Biol. 39: 273,1995

21. Couble ML, Farges JC, Bleicher F, Perrat-Mabillon B, Boudeulle M, Magloire H. Odontoblast differentiation of human dental pulp cells in explant cultures. Calcif Tissue Int. 66: 129,2000

22. Banes AJ, Link GW Jr., Gilbert JW, Tran Son Tay R, Monbureau O. Culturing cells in a mechanically active environment. Am Biotechnol Lab. 8: 12,1990

23. Wang N, Butler JP, Ingber DE. Mechanotransduction across the cell surface and through the cytoskeleton. Science. 260: 1124,1993

24. Kapur S, Baylink DJ, Lau KH. Fluid flow shear stress stimulates human osteoblast proliferation and differentiation through multiple interacting and competing signal transduction pathways. Bone. 32: 241,2003

25. Waldman SD, Spiteri CG, Grynpas MD, Pilliar RM, Kandel RA. Long-term intermittent shear deformation improves the quality of cartilaginous tissue formed *in vitro*. J Orthop Res. 21: 590,2003

26. Owan I, Burr DB, Turner CH, Qiu J, Tu Y, Onyia JE, et al. Mechanotransduction in bone: osteoblasts are more responsive to fluid forces than mechanical strain. Am J Physiol. 273: C810,1997

27. Wadhwa S, Godwin SL, Peterson DR, Epstein MA, Raisz LG, Pilbeam CC. Fluid flow induction of cyclo-oxygenase 2 gene expression in osteoblasts is dependent on an extracellular signal-regulated kinase signaling pathway. J Bone Miner Res. 17: 266,2002

28. Pavasant P, Yongchaitrakul T, Pattamapun K, Arksornnukit M. The synergistic effect of TGF-beta and 1,25-dihydroxyvitamin D3 on SPARC synthesis and alkaline phosphatase activity in human pulp fibroblasts. Arch Oral Biol. 48: 717,2003

29. Alliot-Licht B, Bluteau G, Magne D, Lopez-Cazaux S, Lieubeau B, Daculsi G, et al. Dexamethasone stimulates differentiation of odontoblast-like cells in human dental pulp cultures. Cell Tissue Res. 321: 391,2005

(Honda MJ, Sumita Y, Kagami H, Shinohara Y, Tonomura A, Ohara T, Tsuchiya S, Sagara H, Ueda M)

Chapter 6

Cardiovascular

Bioreactor system for tissue-engineered vascular construct

Although tissue-engineered great vessels and heart valves are currently undergoing clinical trials in some institutions [1, 2], their routine clinical application is still underway. One major problem thwarting their wide clinical usage is lack of physical endurance, which limits their application to cardiovascular structures exposed to high systemic pressure. Although durable scaffold materials have been designed to resist such high pressure *in vivo*, these materials remain subject to degradation over a longer period, resulting in unsatisfactory outcomes after implantation. It is well known that cardiovascular cells such as endothelial or smooth muscle cells are influenced by their extracellular environment, especially local fluid dynamics. Numerous scientists have reported the effect of shear stress or stretch stress on both endothelial and vascular smooth muscle cells [3–5]. Bioreactors for cardiovascular conduits are designed to utilize the effects of those physical strains on the cells to create more optimal tissue for grafting. Although a biomimetic *in vitro* environment is known to increase the endurance of tissue-engineered cardiovascular components [6], the optimal culture conditions including various pressure profiles are not known for a bioreactor. It was our aim to design a bioreactor system that can reproduce a wide range of pulsatile flows with a completely physiological pressure profile. To develop this novel bioreactor system, an intra-aortic balloon pumping (IABP) system was combined with an outflow valve and a compliance chamber to obtain both physiological systolic and diastolic pressures. The compliance chamber and the resistant clamps were designed to reproduce a physiological and relatively wide pressure waveform instead of the original peaky waveform generated by the power source. With a computed manipulation system, this novel bioreactor allows adjustment over a varying range of pressures, pulse rates, and intervals. In this study, we also demonstrated the morphology and the biochemical properties of the tissue-engineered products to illustrate the practicality of this novel bioreactor system. Details of the bioreactor system are presented below.

Bioreactor design

Our bioreactor housing and tubing are made of acryl and polyvinyl chloride (PVC), respectively. The bioreactor is small enough to fit inside a standard CO_2 incubator (Fig. 32), and the driving system is situated outside the incubator. The bioreactor consists of four chambers: a balloon chamber (1), a compliance chamber (2), a culture chamber (3), and a reservoir (4) (Fig. 33). The pressure generated by the IABP is conducted via the balloon (30 or 40 cm³), which is located inside the balloon chamber. The pulsatile flow is generated in the compliance chamber through a one-way outflow valve. The compliance chamber and the culture chamber are connected by a PVC tube with a clamp that controls the fluid flow and pressure. The tissue-engineered products are fabricated inside the culture chamber, which is connected to the reservoir via a PVC tube. A resistance clamp is located in the middle of the connection tube to regulate the afterload. The air filter on the reservoir regulates the amount of CO_2 in the atmosphere. Fluids recirculate back to the balloon chamber via one-way valve from the reservoir. This system is sterilized with ethylene oxide. Conditions in the culture chamber are monitored by a pressure meter and a flow meter, and are recorded by a computerized analyzer.

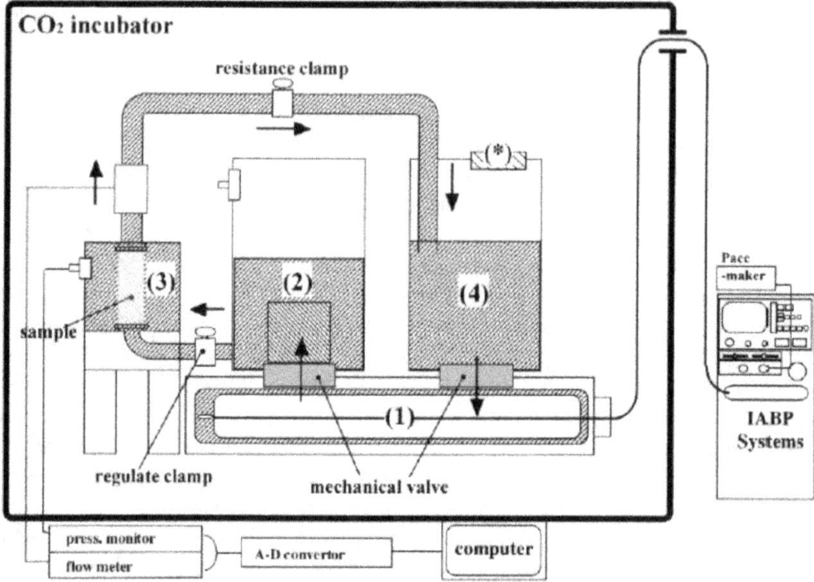

Fig. 32. Schema of the pulse-duplicating bioreactor system. The bioreactor is small enough to fit inside a standard CO_2 incubator, and the driving system is situated outside the incubator. The basal part of this bioreactor is the balloon chamber (1). Control of inflow to and outflow from the balloon chamber is by a mechanical valve. The compliance chamber (2) buffers the pumping flow and pressure. The engineered tissue sample is located in the culture chamber (3). A regulating clamp is placed between (2) and (3). The next chamber is the reservoir equipped with an air filter for a CO_2 incubator (4). The air filter (*) regulates the amount of CO_2 in the atmosphere. A resistance clamp between (3) and (4) regulates the afterload. Conditions in the culture chamber are monitored by a pressure meter and a flow meter, and are recorded by a computerized analyzer (From Narita et al. 2004. Reprinted with permission).

Bioreactor function analyses
After confirming the capability of this bioreactor system for flow, pressure, and frequency, we tried to evaluate the actual physical strains for the culture cells. First, the shear stress level (γ) in the scaffold was calculated as

$$\gamma = \mu \times du/dr = 4\mu Q/\pi r^3$$

where μ is the viscosity of the culture medium, du/dr is the velocity gradient, Q is the flow rate of the bioreactor, and r is the radius of the scaffold under the physiological or peaky waveform pattern of pulsatile flow (with a mean flow rate of 500 ml/min and 40 mmHg of systolic pressure). Next, we estimated the uniaxial cyclic strain under pulsatile flow by measuring the diameter change of the scaffold under various pressure conditions, using a specially designed system. The system consisted of a digital video camera, a pressure monitor, a flow meter, and an analyzing computer. Under various conditions, the outside diameter of the scaffold, which was mounted in the culture chamber, was recorded with a digital video camera. The images were downloaded to a computer and the diameter was measured with NIH Image software. With this system, the actual stretch of the scaffold (only the direction perpendicular to the axis of the scaffold) could be estimated as a change in the scaffold radius as follows:

$$\partial = 100 \times (\pi D\text{max} - \pi D\text{min}) / \pi D\text{min}$$

where ∂ is the percentage of radial diameter change (which is equivalent to the change of scaffold length around the axis), Dmax is the maximum diameter, and Dmin is the minimum diameter of the scaffold. The sensitivity of the system was enough to detect a difference of more than 0.2%.

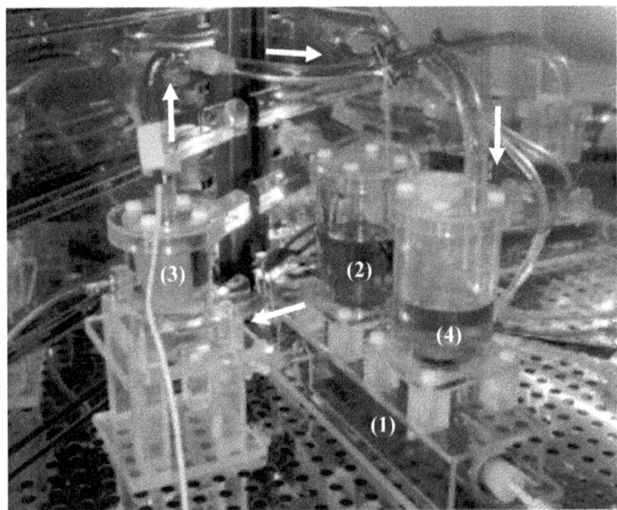

Fig. 33. Experimental setting of the pulse-duplicating bioreactor. The bioreactor consists of four chambers: a balloon chamber (1), compliance chamber (2), culture chamber (3), and reservoir (4). Pulsatile flow is generated in the compliance chamber (2) through a one-way outflow valve. The compliance chamber (2) and the culture chamber (3) are connected by a PVC tube with a clamp that controls the fluid flow and pressure. Tissue-engineered products are fabricated inside the culture chamber (3), which is connected to the reservoir (4) via a PVC tube. Fluids recirculate back to the balloon chamber (1) via a one-way valve from the reservoir (4) (From Narita et al. 2004. Reprinted with permission).

The combination of an outflow valve, compliance chamber, and resistant clamps together with a balloon pumping system was able to successfully reproduce both physiological systolic and diastolic pressures. The compliance chamber was especially effective in transforming the original peaky pressure waveform into a physiological pressure profile. The tissues, cultured under a physiological pressure waveform with pulsatile flow, presented widely distributed cells in close contact with each other. They also showed significantly higher cell numbers, total protein content, and proteoglycan–glycosaminoglycan content than cultured tissues under a peaky pressure wave or under static conditions (Fig. 34). This new bioreactor system is suitable for evaluating a favorable environment for tissue-engineered cardiovascular components.

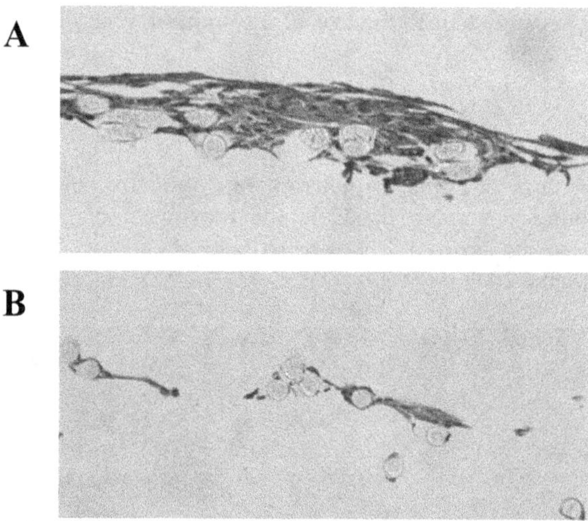

Fig. 34. Bright-field photomicrographs showing H-E staining of newly formed tissues under dynamic and static conditions. A: The tissues in group D-w were exposed to pressure with physiological wide-shaped waveforms for 7 days. B: Tissue in group S was maintained under static conditions. Note that cell density was clearly higher under dynamic conditions than under static conditions (Original magnification × 100) (From Narita et al. 2004. Reprinted with permission).

References

1. Shin'oka T, Imai Y, Ikada Y. Transplantation of a tissue-engineered pulmonary artery. N Engl J Med. 344: 532,2001
2. Dohmen PM, Lembcke A, Hotz H, Kivelitz D, Konertz WF. Ross operation with a tissue-engineered heart valve. Ann Thorac Surg. 74: 1438,2002
3. Ando J, Tsuboi H, Korenaga R, Takada Y, Sorimachi N, Miyasaka M, Kamiya A. Shear stress inhibits adhesion of cultured mouse endothelial cells to lymphocytes by downregulating VCAM-1 expression. Am J Physiol. 267: C679,1994
4. Naruse K, Sokabe M. Involvement of stretch-activated ion channels in Ca2+ mobilization to mechanical stretch in endothelial cells. Am J Physiol. 264: C1037,1993
5. Seliktar D, Black RA, Vito RP, Nerem R. Dynamic mechanical conditioning of collagen–gel blood vessel constructs induces remodeling *in vitro*. Ann Biomed Eng. 28: 351,2000
6. Hoerstrup SP, Sodian R, Daebritz S, Wang J, Bacha EA, Martin DP, Moran AM, Guleserian KJ, Sperling JS, Kaushal S, Vacanti JP, Schoen FJ, Mayer JE. Functional living trileaflet heart valves grown *in vitro*. Circulation. 102(Suppl.III): 44,2000

(Narita Y, Hata K, Kagami H, Usui A, Ueda Y, Ueda M)

Chapter 7

Ureter

Tissue-engineered ureter using a decellularized matrix

Partial ureterectomy is performed for patients with ureteral cancer or severe retroperitoneal fibrosis [1, 2]. When the length of the dissected ureter is long, especially proximally, ureteral replacement is required to avoid urinary diversion, such as nephrostomy or autorenal transplantation. However, previous trials using artificial materials have failed due to infection, inflammation, or calcification [3-7]. To improve the quality of life for patients undergoing nephrostomy, development of a novel graft material is needed. Decellularized matrices have been widely used as scaffolds for tissue engineering. Compared with biodegradable polymers, decellularized matrices possess physiological properties that are closer to those of normal tissue. These properties are beneficial for forming tubular structures, such as blood vessels, with similar pliancy to that of normal vessels [8]. Furthermore, decellularized matrices are highly biocompatible, which facilitates reepithelialization after transplantation [9]. This is of great benefit when a scaffold requires an epithelial lining. Decellularized matrices are also known to be minimally immunogenic, even in the case of xenografts, which is advantageous when a stable supply of scaffold material is required for clinical applications [10, 11]. One of the most important factors for successful tissue-engineered ureters is thought to be a functional lining with uroepithelial cells (UECs), which are expected to maintain the luminal space, prevent calculus formation, and possibly resist stricture of the duct [3-6]. Fortunately, UECs from the bladder are readily available clinically, by endoscopy, without radical surgical procedures. Thus, the combination of an allogenic or xenogeneic decellularized ureter with autologous UECs might generate an ideal material for grafting. Decellularized matrices have been used to construct tissue-engineered urinary bladders and ureters in rat models [9-12]. However, replacement of the ureter in larger animals has not yet been successful [13]. We suggest that one of the major obstacles to application with the large animal models is providing a blood supply for the grafted epithelial cells, as the thickness of the scaffold in animals such as dogs is above the limit of diffusion for nutrients and oxygen from the surrounding tissue, causing the grafted cells to starve. Unlike tissue-engineered blood vessels, tissue-engineered ureters cannot access a blood supply until neovascularization is completed. In fact, our preliminary experiments have shown that epithelial cells seeded onto canine ureteral decellularized matrices (UDMs) gradually disappear over the 14-day period after transplantation. Several approaches have been developed in an attempt to improve the blood supply for the transplanted tissue, including the application of growth factors [14, 15]. Bone marrow-derived mononuclear cells (BM-MNCs) are known to contain endothelial progenitor cells and secrete several growth factors [16]. They have also been reported to accelerate neovascularization in ischemic diseases, such as Burger's disease and ischemic heart disease, and might constitute a novel strategy for accelerating angiogenesis [17-19]. We were interested in whether BM-MNCs could facilitate neovascularization of a transplanted decellularized scaffold. The aim of the present study was to investigate the potential of the decellularized ureter as a scaffold for constructing a tissue-engineered ureter and the role of BM-MNCs in enhancing angiogenesis in the UDM to enable the seeded UECs to survive. Below is a detailed explanation of the method.

Generation of UDM

To render the canine ureteral matrices decellular, they were excised from the beagles and treated with deoxycholic acid, as reported previously [9]. Briefly, the ureters were shaken in 10 ml of 9.1 mM PBS containing 0.1% sodium azide for 12 hours in a humidified 5% CO_2 atmosphere at 37°C. Next, the samples were treated with 10 ml 0.1 M sodium chloride containing 0.2 mg/ml deoxyribonuclease for 12 hours, followed by 10 ml PBS containing 4% sodium deoxycholate and 0.1% sodium azide for 24 hours. The treated tissues were evaluated histologically to confirm their decellular status and assess damage to the extracellular matrices. The tissues were stored in PBS with 1% antibiotic-antimycotic solution at 4°C.

Cell seeding on UDM surface

Sections (2 cm) of the UDM were perfused with 5% fibronectin in PBS for 24 hours at 4°C in order to improve cell attachment. One end of the UDM was ligated, the inner space was filled with 1×10^6 cultured UECs, and the other end of the duct was ligated. The UDM was placed in a tube with culture medium and rotated for 8 hours. The ligatures were removed and the UDM with cells was incubated in a flask for a further 3 days before transplantation.

Transplantation of UDMs with UECs: Experiment 1

As a preliminary experiment, UDMs seeded with UECs (UDM-UECs) were transplanted to the omentum of nude rats in order to investigate the fate of the seeded cells. At the time of surgery, the animals were anesthetized with sodium pentobarbital (50 mg/kg body weight injected intraperitoneally). The samples were removed and evaluated histologically at 3 days (n = 3) and 14 days (n = 3) after transplantation.

BM-MNC seeding onto UDM-UEC: Experiment 2

Bone marrow (10 ml) was aspirated from the iliac crests of dogs anesthetized with intravenous pentobarbital. BM-MNCs were isolated using a density-gradient method [19], according to the manufacturer's protocol, and suspended at 1×10^6 cells/ml in DMEM. UDM-UECs were randomly divided into two groups. In group A (n = 3), the UDM-UECs were implanted into the subcutaneous space of nude mice without BM-MNCs, as controls. In group B (n = 3), BM-MNCs (3×10^5 cells in 300 µl) were seeded onto the outer surface of the UDM-UECs before transplantation. The grafts were removed and evaluated after 14 days. The effect of BM-MNCs on neovascularization in the grafts was assessed in two ways. First, the density of microvessels was measured by immunostaining with an antifactor VIII antibody by counting the number of microvessels in at least three fields (0.74 mm²) and calculating the mean number. Second, the distance from the inner surface of the UDM to the closest capillary was measured at 10 randomly selected points in each graft, and the mean values were calculated.

Combined effects of seeded BM-MNCs and the recipient site: Experiment 3

To investigate the effects of the recipient site on neovascularization, UDM-UECs with BM-MNCs were transplanted into the omentum of nude rats and compared with those transplanted into the subcutaneous space of nude mice. The grafts were removed after 14 days and processed for histological evaluation.

The cultured UECs began showing vacuolar degeneration 3 days after transplantation and gradually disappeared thereafter. To facilitate neovascularization in the implant, BM-MNCs were seeded around the UDM before transplantation. This facilitated the survival of the UECs, which formed three to five cellular layers after 14 days. The mean microvessel density was significantly increased in tissues seeded with BM-MNCs. However, cell-tracking experiments revealed that the increased number of capillaries in the experimental group was not due to the direct differentiation of transplanted endothelial progenitor cells. Our results demonstrate that the UDM is a useful scaffold for a tissue-engineered ureter, especially when seeded with BM-MNCs to enhance angiogenesis (Figs. 35, 36).

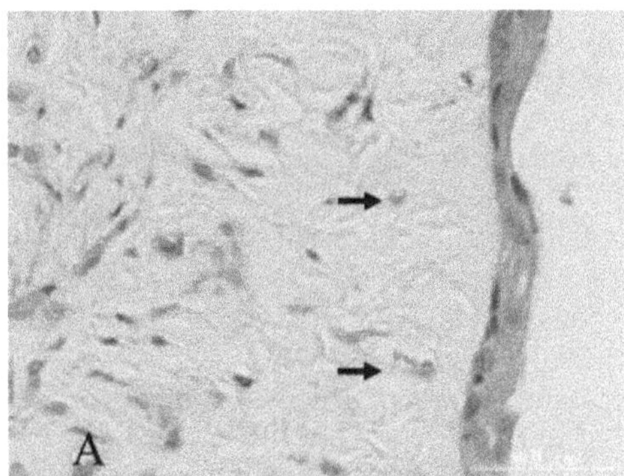
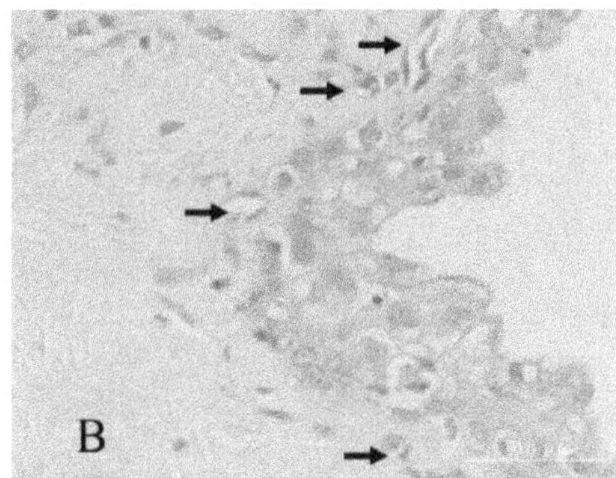

Fig. 35. Bright-field photomicrographs showing H-E staining of tissue sections from UDM-UECs transplanted with BM-MNCs to nude mouse subcutaneous space (Experiment 2) (A) and to nude rat omentum (Experiment 3) (B). Histological section from UDM-UECs with BM-MNCs implanted into the omentum exhibited a thicker epithelial cell layer than these implanted into subcutaneous space. Samples were evaluated 14 day after transplantation. In both sections, capillaries were observed close to UECs (arrows) (From Matsunuma et al. 2006. Reprinted with permission).

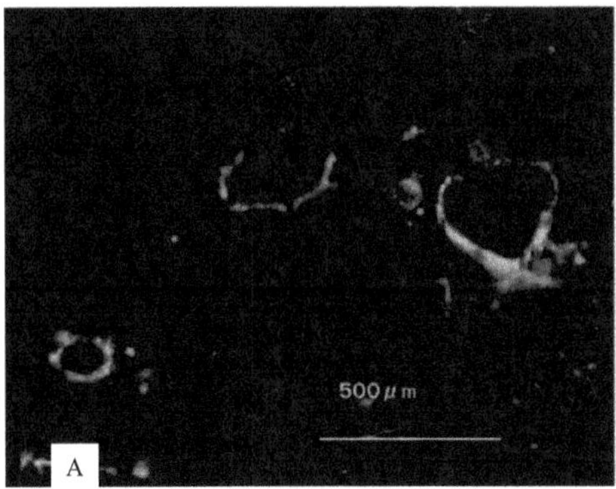
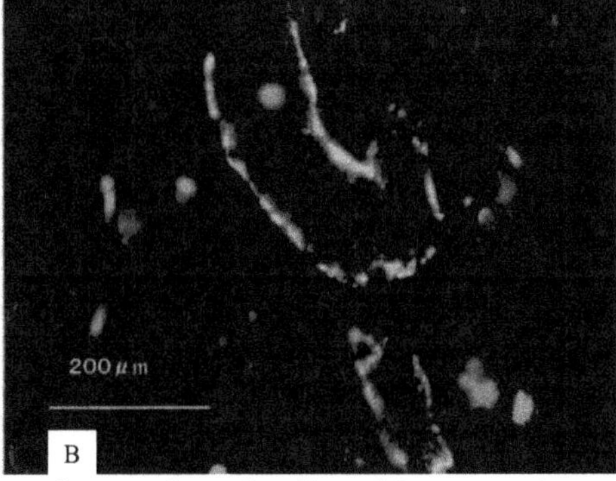

Fig. 36. Dark-field photomicrographs showing the results of cell-tracking experiments. BM-MNCs were labeled with the red fluorescent cell dye, PKH26, before being seeded onto implants, and sections of implants after 14 days were labeled for factor VIII in green using an FITC-labeled antibody. Note that only a few cells were double-labeled. A: Original magnification × 100; B: × 200 (From Matsunuma et al. 2006. Reprinted with permission).

This study of the initial establishment of grafts *in vivo* showed that UDMs were highly biocompatible with cultured UECs and should be a useful scaffold material for tissue-engineered ureters. Preseeding the grafts with BM-MNCs was a key factor in promoting the survival of the UECs by accelerating cell migration and neovascularization.

References

1. Pohar KS, Sheinfeld J. When is partial ureterectomy acceptable for transitional-cell carcinoma of the ureter. J Endourol. 15: 405,2001
2. Katz R, Golijanin D, Pode D, Shapiro A. Primary and postoperative retroperitoneal fibrosis-experience with 18 cases. Urology. 60: 780,2002
3. Warren JW Jr., Coomer T, Fransen H. The use of teflon grafts for replacement of ureters. J Urol. 89: 164,1963
4. Ulm AH. Total replacement of the human ureter with teflon prosthesis. J Urol. 96: 455,1966
5. Homann, W., Schreiber, B., Mlynek, M.L., Kropfl, D., and Fenkl, H.F. Long-term results of prosthetic ureteral replacement in minipigs. Urol Int. 39: 95,1984
6. Jonas D, Tschirkov F, Krause E. Autologous collagen graft for partial replacement of the ureter (author's translation). Urol Int. 28: 290,1973
7. Ahmadzadeh M. Clinical experience with subcutaneous urinary diversion: new approach using a double pigtail stent. Br J Urol. 67: 596,1991
8. Schaner PJ, Martin ND, Tulenko TN, Shapiro IM, Tarola NA, Leichter RF, Carabasi RA, Dimuzio PJ. Decellularized vein as a potential scaffold for vascular tissue engineering. J Vasc Surg. 40: 146,2004
9. Cayan S, Chermansky C, Schlote N, Sekido N, Nunes L, Dahiya R, Tanagho EA. The bladder acellular matrix graft in a rat chemical cystitis model: functional and histologic evaluation. J Urol. 168: 798,2002
10. Probst M, Piechota HJ, Dahiya R, Tanagho EA. Homologous bladder augmentation in dog with the bladder acellular matrix graft. BJU Int. 85: 362,2000
11. Elkins RC, Dawson PE, Goldstein S, Walsh SP., and Black, K.S. Decellularized human valve allografts. Ann. Thorac. Surg. 71: S428,2001
12. Dahms SE, Piechota HJ, Nunes L, Dahiya R, Lue TF, Tanagho EA. Free ureteral replacement in rats: regeneration of ureteral wall components in the acellular matrix graft. Urology. 50: 818,1997
13. Shalhav AL, Elbahnasy AM, Bercowsky E, Kovacs G, Brewer A, Maxwell KL. McDougall, E.M., and Clayman, R.V. Laparoscopic replacement of urinary tract segments using biodegradable materials in a large-animal model. J Endourol. 13: 241,1999
14. Kanematsu A, Yamamoto S, Noguchi T, Ozeki M, Tabata Y, Ogawa O. Bladder regeneration by bladder acellular matrix combined with sustained release of exogenous growth factor. J Urol. 170: 1633,2003
15. Giulian D, Woodward J, Young DG, Krebs JF, Lachman LB. Interleukin-1 injected into mammalian brain stimulates astrogliosis and neovascularization. J Neurosci. 8: 2485,1988
16. Kaigler D, Krebsbach PH, Polverini PJ, Mooney DJ. Role of vascular endothelial growth factor in bone marrow stromal cell modulation of endothelial cells. Tissue Eng. 9: 95,2003

17. Kamihata H, Matsubara H, Nishiue, T, Fujiyama S, Tsutsumi Y, Ozono R, Masaki H, Mori Y, Iba O, Tateishi E, Kosaki A, Shintani S, Murohara T, Imaizumi T, Iwasaka T. Implantation of bone marrow mononuclear cells into ischemic myocardium enhances collateral perfusion and regional function via side supply of angioblasts, angiogenic ligands, and cytokines. Circulation. 104: 1046,2001

18. Kawamoto A, Gwon HC, Iwaguro H, Yamaguchi JI, Uchida S, Masuda H, Silver M, Ma H, Kearney M, Isner JM, Asahara T. Therapeutic potential of *ex vivo* expanded endothelial progenitor cells for myocardial ischemia. Circulation. 103: 634,2001

19. Orlic D, Kajstura J, Chimenti S, Jakoniuk I, Anderson SM, Li B, Pickel J, McKay R, Nadal-Ginard B, Bodine DM, Leri A, Anversa P. Bone marrow cells regenerate infarcted myocardium. Nature. 410: 701,2001

(Matsunuma H, Kagami H, Narita Y, Hata K, Ono Y, Ohshima S, Ueda M)

Chapter 8

Salivary gland

Selected salivary grand cell culture

Salivary gland atrophy is commonly seen in patients in whom those salivary glands have been irradiated for oral and facial cancer treatment. In addition, aging has been reported to induce histological changes in the parenchyma of the salivary glands such as the replacement with fat or connective tissue [1]. The atrophic gland leads to dry mouth, a condition frequently associated with various symptoms including active dental caries, a burning sensation in the mouth, and difficulty with eating, swallowing or speech. Moreover, an increased incidence of infection such as oral candidiasis has been reported [2]. Patients with dry mouth have been treated with salivary substitutes and/or medications such as pilocarpine or cevimeline hydrochloride [3]. These treatments temporarily relieve the patients' symptoms, inducing salivation from the residual tissue. However, no treatment is available for the purpose of regenerating the atrophic glands.

Recently, the possibility of re-engineering atrophic or damaged salivary glands has been explored with the aim of developing novel clinical treatments [4]. Gene therapy for the salivary gland shows promise as a future treatment adjunct, since gene transfer to an irradiated salivary gland has been shown to increase fluid secretion in an animal model [5]. However, the clinical application of gene therapy to the atrophic glands requires further development and validation of the new vehicles for gene delivery, since currently available viral or nonviral vectors still require some considerations regarding safety, efficiency or the duration of transgene expression. Tissue engineering is a promising new approach to the replacement of lost or damaged tissue. Therefore, we attempted to establish a method for culturing normal epithelial salivary gland cells. Monolayer culture of salivary-gland cells has the advantage of allowing investigation of the effects of individual biochemical substances without the influence of other modifying factors. However, the use of monolayer culture techniques on cells from the normal salivary gland requires the elimination of contaminating fibroblasts and the maintenance of cell proliferation for prolonged periods. To avoid contamination by fibroblasts and other cell types, Olivar tried to select specific cell types using Ficoll (Pharmacia) density-gradient centrifugation, but he reported only the suspension culture protocols [6]. The same aims were addressed by Yang et al., using low Ca^{2+} concentrations in his medium as well as a collagen gel matrix [7]. This method exhibited fibroblast proliferation and allowed salivary-gland cells to exhibit a duct-like appearance. While this technique permitted investigation of tissue morphology, it was difficult to investigate cellular proliferation. To inhibit contamination by fibroblasts, and to obtain a stable monolayer of cultured cells in a short time, Sabatini et al. used 3T3 cells as a feeder layer for salivary-gland cell culture [8]. The 3T3 cells inhibited the proliferation of fibroblasts while maintaining the proliferative activity of adjacent epithelial cells for prolonged periods of time without differentiation. The same group reported the presence of two types of cell in these cultures; one cuboidal and in close contact with the surrounding cells, the other rounded without cell-to-cell contact. So, we modified their method to selectively culture acinar cells during the first and second passages. The method is described in detail in the following

Preparation of 3T3 cells
The 3T3 cells were kindly provided by Dr. Howard Green (Department of Cellular and Molecular Physiology, Harvard Medical School, Boston, MA, U.S.A.). To suppress cell proliferation, they were treated with 4 µg/ml mitomycin C in DMEM for 2 hours at 37°C without fetal bovine serum. Cells treated in this manner promote the proliferation of adjacent epithelial cells but do not proliferate themselves. After incubation, the 3T3 cells were rinsed three times with Hank's solution to remove the mitomycin C.

Tissue preparation and salivary gland epithelial cell culture
The submandibular glands were excised from Wistar rat and rinsed in PBS containing 100 U/ml penicillin, 0.1 mg/ml streptomycin, and 0.25 µg/ml amphotericin B. Using curved scissors, the tissues were cut into cubes of at least 1 mm³. Salivary gland cells were dissociated by digestion with a solution of 0.1 mg/ml collagenase in PBS for 30 minutes at room temperature. Cells were then stirred for 30 minutes with a magnetic stirring bar, filtered through a 70-µm nylon mesh, and collected by centrifugation for 5 minutes at 1500 rpm. The separated salivary gland cells were cultured in 75-cm plastic flasks, in a humidified atmosphere of 10% CO_2 at 37°C in a 3: 1 mixture of DMEM and Ham's F-12 containing 5% fetal bovine serum, 5.0 µg/ml insulin, 5.0 µg/ml transferrin, 2×10^{-9} M tri-iodothyronine, 1×10^{-9} M cholera toxin, 0.4 µg/ml hydrocortisone, 0.1 µg/ml epidermal growth factor, 10 U/ml penicillin, 0.1 mg/ml streptomycin, and 0.25 µg/ml amphotericin B (Fig. 37). To confirm the origin of these cultured cells, amylase production was examined by electron microscopy and periodic acid Schiff staining, together with immunocytochemical analysis of myosin, anticytokeratin (CK-1, CK 10/13, CK-MNF116, CK-LMW, CK-HMW and CK-19) and amylase antibody.

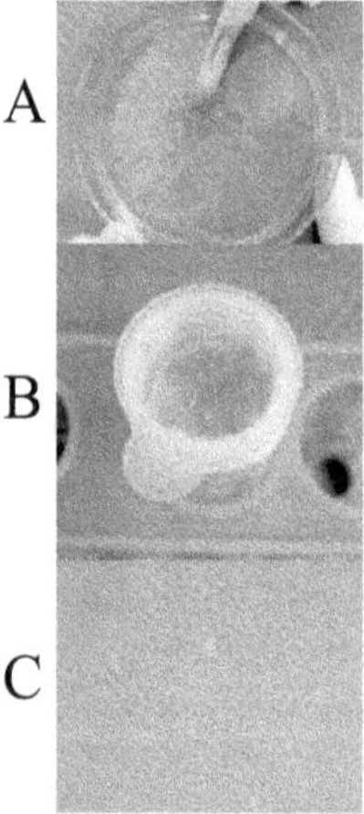

Fig. 37. Protocol of salivary gland epithelial cell culture. A: Excise salivary glands, and cut into cubes of at least 1 mm³. B: Dissociation by digestion with a solution of 0.1 mg/ml collagenase in PBS for 30 minutes and filtered through a 70-µm nylon mesh. C: Cell culture (From Horie et al. 1996. Reprinted with permission).

The cultured cells demonstrated secretion granules containing amylase and presented features characteristic of acinar cells, which they retained until passage two. By using a feeder layer in conjunction with a newly formulated culture medium, the selectability of these cells was improved. Changes in proliferation of cultured salivary-gland cells in the presence of selected neurotransmitters were also examined, Isoproterenol enhanced cellular proliferation.

Thus, we established a culture method for normal epithelial salivary gland cells. However, tissue engineering of a whole organ is still not feasible since it requires the regeneration of a complex ductal system, blood supply and renervation, none of which are possible with the current technology. Cell transplantation is also a possible treatment option to restore the function of lost or damaged tissue, and has been extensively studied in cases of type I diabetes [9-12], Parkinson's disease [13-15], and liver dysfunction [16]. Most such studies have aimed to restore the function of a specific cell type or cell groups in an organ. The possibility of regenerating an entire organ by cell transplantation has not been demonstrated except for the liver, which possesses enormous regenerative capacity [17]. Recent studies have shown that transplanted bone marrow cells might replace tissue in many more organs than was previously thought possible. If this is the case, transplanted salivary gland stem cells could eventually restore at least part of the organ after extensive cell proliferation and differentiation. Since the cell therapy approach can avoid some of the technical difficulties posed by tissue engineering, it is worthwhile investigating its potential for the treatment of atrophic salivary glands.

Cell transplantation into an atrophic salivary gland

Next, we investigated the feasibility of a cell therapy for an atrophic salivary gland using cultured salivary gland cells, we established the culture method. First, we investigated the survival and distribution of transplanted cells in the salivary gland. Second, we demonstrated a possible differentiation of the transplanted cells. Third, we examined the possible drawbacks of cell transplantation on normal salivary glands since the cells could be harmful if they attach and remain within a normal tissue structure. Below is how to generate an atrophic salivary gland model and transplant the cultured cells into atrophic glands.

Generation of atrophic submandibular gland by ductal ligation

The atrophic submandibular gland was generated by a ductal ligation according to Tamarin with modifications [18, 19]. Briefly, the main ducts of the left submandibular-sublingual glands of male Wistar rats (each weighing about 200 g) were exposed surgically. The ducts with some surrounding connective tissue investments were isolated by blunt dissection under a surgical microscope. A specially prepared stainless steel tube and a nylon ligature were looped around each duct and tied. The tissues were then reapposed and the skin was sutured. The ligation period lasted one week for all experiments. One week after ductal ligation, atrophy of the acinar cells was observed. Ducts were dilated and fibrous connective tissues were prominent.

Cell transplantation

The procedure and protocol for our cultured salivary gland epithelial cell transplantation are summarized in figure (Fig. 38). Prior to transplantation, the cultured salivary gland epithelial cells were detached from culture dishes by enzymatic treatment with 0.05% trypsin-EDTA. The cells were then labeled with a fluorescent cell linker PKH26 at a concentration of 2×10^{-6} M for 5 minutes. The cell pellet was resuspended in DMEM containing 10% fetal bovine serum to yield a density of 50,000 cells/µl. India ink was added to the cell suspension at a 1:200 dilution in order to locate the injection site. Twenty microlitters of the cell suspension containing 2×10^6 PKH26-labeled cultured cells was injected through a 25-gauge needle into the atrophic submandibular gland. When the injections were properly placed, India ink was visible below the capsules. In the control group, the same cell suspension was injected into normal submandibular glands.

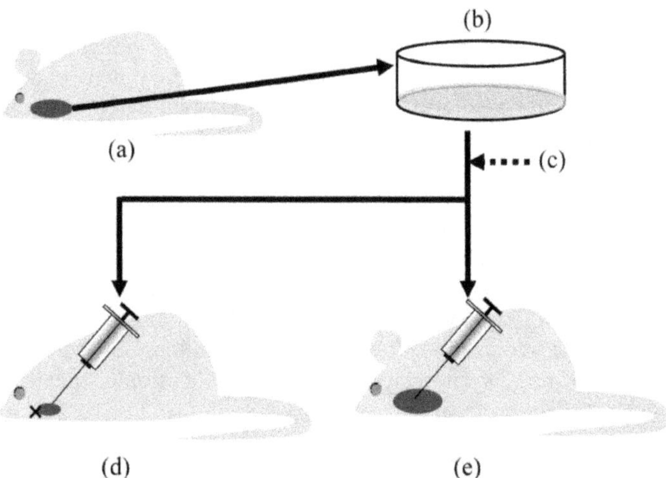

Fig. 38. Schema ofexperimental protocol. (a) Excise submandibular glands. (b) Cell culture. (c) Cell labeling with PKH. (d) Cell transplantation intoatrophic submandibular glands. (e) Cell transplantation into normal submandibular glands (From Sugito et al. 2004).

The submandibular glands were excised at 2 and 4 weeks after transplantation. At least four animals were used at each of the above time points in both the experimental and control groups. The excised glands were fixed in 10% buffered formalin and embedded in OCT Compound on dry ice. The samples were then stored at -80°C until the time for cryosectioning. Two weeks after cell transplantation, the transplanted cells were detectable in both the experimental and control groups. The cells were clustered in the connective tissue between the lobules. Four weeks after transplantation, the labeled cells were detectable in the experimental group but not in the control group. In the atrophic glands, the scattered transplanted cells were observed over a broad area of the gland but localized mainly around the acini and ductal region. Immunostaining results showed a possible involvement of the transplanted cells in ductal regeneration, while neither myoepithelial nor acinar differentiations were observed within the four weeks since transplantation. This study demonstrated that cell transplantation to the salivary gland is feasible, and that the transplanted cells were selectively attracted to and remained in the damaged area without affecting normal tissue.
The present study described the possibility of cell therapy for the salivary gland by use of cultured normal salivary epithelial cells. However, these results are very preliminary since the period of observation was short, and the longer-term fate of the transplanted cells is not known. Long-term observations as well as investigations into a suitable cell source are areas for future studies.

References

1. Sreebny LM, Edgar WM, O'Mullane DM. Xerostomia: diagnosis, management and clinical complications. In Saliva and oral health, Second edition. London, UK: British Dental Association. 45,1996
2. Ernesto A, Soto-Rojas, Kraus A. The oral side of Sjogren syndrome. Diagnosis and treatment. A Review. Arch Med Res. 33: 95,2002
3. Fife RS, Chase WF, Dore R K, Wiesenhutter CW, Lockhart PB, Tindall E, Suen JY. Cevimeline for the treatment of xerostomia in patients with Sjogren syndrome: a randomized trial. Arch Intern Med. 162: 1293,2002
4. Baum BJ. Prospects for reengineering salivary glands. Adv. Dent. Res. 14: 84,2000
5. Delporte C, O'Connell BC, He X, Lancaster H E, O'Connell AC, Agre P, Baum BJ. Increased fluid secretion after adenoviral-mediated transfer of the aquaporin-1 cDNA to irradiated rat salivary glands. Proc Natl Acad Sci U S A. 94: 3268,1997
6. Olivar C. Isolation and maintenance of differentiation exocrine gland acinar cells *in vitro*. In *vitro*. 16: 297,1980
7. Yang J, Flynn D, Larson L, Hamamoto S. Growth in primary culture of mouse submandibular epithelial cells embedded in collagen gels. *In vitro*. 18: 435,1982
8. Sabatini LM, A-Hoffmann BL, Warner TF, Azen EA. Serial cultivation of epithelial cells from human and macaque salivary glands. *In vitro* cell dev Biol. 27: 939,1991
9. Moore WV, Bieser K, Geng Z, Tong PY, Kover K. Decreased survival of islet allografts in rats with advanced chronic complications of diabetes. Cell Transplant. 11: 707,2002
10. Ricordi C, Finke EH, Dye ES, Socci C, Lacy PE. Automated isolation of mouse pancreatic islets. Transplantation. 46: 455,1988
11. Shapiro AM, Lakey JR, Ryan EA, Korbutt GS, Toth E, Warnock GL, Kneteman N, M, Rajotte RV. Islet transplantation in seven patients with type 1 diabetes mellitus using a glucocorticoid-free immunosuppressive regimen. N Engl J Med. 343: 230,2000
12. Shibata A, Ludvigsen CW Jr., Naber SP, McDaniel ML, Lacy PE. Standardization for a digestion-filtration method for isolation of pancreatic islets. Diabetes. 25: 667,1976
13. Bjorklund AU. Reconstruction of the nigrostriatal dopamine pathway by intracerebral nigral transplants. Brain Res. 177: 555,1979
14. Lindvall O, Rehncrona S, Brundin P, Gustavii B, Astedt B, Widner H, Lindholm T, Bjorklund A, Leenders KL, Rothwell JC. Human fetal dopamine neurons grafted into the striatum in two patients with severe Parkinson's disease. A detailed account of methodology and a 6-month follow-up. Arch Neurol. 46: 615,1989
15. Perlow MJ, Freed WJ, Hoffer B J, Seiger A, Olson L, Wyatt R J. Brain grafts reduce motor abnormalities produced by destruction of nigrostriatal dopamine system. Science. 204: 643,1979
16. Susick R, Moss N, Kubota H, Lecluyse E, Hamilton G, Luntz T, Ludlow J, Fair J, Gerber D, Bergstrand K, White J, Bruce A, Drury O, Gupta S, Reid LM. Hepatic progenitors and strategies for liver cell therapies. Ann NY Acad Sci. 944: 398,2001
17. Ilan Y, Roy-Chowdhury N, Prakash R, Jona V, Attavar P, Guha C, Tada K. Roy-Chowdhury, J. Massive repopulation of rat liver by transplantation of hepatocytes into specific lobes of the liver and ligation of portal vein branches to other lobes. Transplantation. 64: 8,1997

18. Okazaki Y, Kagami H, Hattori T, Hishida S, Shigetomi T, Ueda M. Acceleration of rat salivary gland tissue repair by basic fibroblast growth factor. Arch Oral Biol. 45: 911,2000
19. Tamarin A. Submaxillary gland recovery from obstruction. I. Overall changes and electron microscopic alterations of granular duct cells. J Ultrastruct Res. 34: 276,1971

(Sugito T, Kagami H, Hata K, Nishiguchi H, Ueda M)

Index

P

Particulate cancellous bone and marrow, 30
Periodontal treatment, 35
Platelet-derived growth factor (PDGF), 4, 34
Platelet-rich plasma (PRP), 22, 31
Periosteal cells, 26
Periosteum, 26
Polyglycolic acid (PGA), 51, 55

R

Regenerative medicine, 1

S

Salibary grand, 71
Scaffolds, 3
Shear stress, 54, 61
Skin, 5
Sinus augmentation, 32
Soft tissue augmentation, 12
Soft tissue management, 11
Stem cells, 2
Syngenic cells, 2

T

Telomerase, 7
Telomere, 6, 7
3T3 cells, 5, 72
Transforming growth factor beta
(TGF-β), 4, 34, 47
Tissue-engineered osteogenic material
(TEOM), 24
Tissue engineering, 1
Tooth bud, 51
Translational research, 29

U

Ultrasound (US), 47
Ureter, 65
Uroepithelial cells, 65

V

Vickers hardness test, 34
Vascular endothelial growth factor
(VEGF), 4, 21

W

Wrinkles, 12

X

Xenogenic cells, 2

www.ingramcontent.com/pod-product-compliance
Lightning Source LLC
LaVergne TN
LVHW060413070726
842759LV00006BA/55